AF548529

Methodologies in Asymmetric Catalysis

ACS SYMPOSIUM SERIES **880**

Methodologies in Asymmetric Catalysis

Sanjay V. Malhotra, Editor
New Jersey Institute of Technology

Sponsored by the
ACS Division of Organic Chemistry

American Chemical Society, Washington, DC

Library of Congress Cataloging-in-Publication Data

Methodologies in asymmetric catalysis / Sanjay V. Malhotra, editor ; sponsored by the ACS Division of Organic Chemistry.

p. cm.—(ACS symposium series ; 880)

Includes bibliographical references and index.

ISBN 0–8412–3835–9 (alk. paper)

1. Asymmetric synthesis—Congresses. 2. Catalysis—Congresses.

I. Malhotra, Sanjay V. II. American Chemical Society. Division of Organic Chemistry. III. American Chemical Society. Meeting (223rd : 2002 : Orlando, Fla.) IV. Series.

D262.M53 2004
547′.2—dc22 2004046177

The paper used in this publication meets the minimum requirements of American National Standard for Information Sciences—Permanence of Paper for Printed Library Materials, ANSI Z39.48–1984.

Distributed by Oxford University Press

PRINTED IN THE UNITED STATES OF AMERICA

Foreword

The ACS Symposium Series was first published in 1974 to provide a mechanism for publishing symposia quickly in book form. The purpose of the series is to publish timely, comprehensive books developed from ACS sponsored symposia based on current scientific research. Occasionally, books are developed from symposia sponsored by other organizations when the topic is of keen interest to the chemistry audience.

Before agreeing to publish a book, the proposed table of contents is reviewed for appropriate and comprehensive coverage and for interest to the audience. Some papers may be excluded to better focus the book; others may be added to provide comprehensiveness. When appropriate, overview or introductory chapters are added. Drafts of chapters are peer-reviewed prior to final acceptance or rejection, and manuscripts are prepared in camera-ready format.

As a rule, only original research papers and original review papers are included in the volumes. Verbatim reproductions of previously published papers are not accepted.

ACS Books Department

Contents

Indexes

Preface

This book is based on papers presented at the American Chemical Society (ACS) national meeting symposium titled *New Methodologies in Asymmetric Catalysis,* held in Orlando, Florida, April 7–11, 2002. Asymmetric catalysis has been one of the important topics of research for scientist in both industrial laboratories and the academic world during the past three decades. This symposium was the first open international meeting on asymmetric catalysis, since the announcement of a Nobel Prize in Chemistry for 2001, which was shared by three scientists (William S. Knowles, Ryoji Noyori, and K. Barry Sharpless) for their pioneering work in this field. The sessions at the ACS symposium on this topic were the best attended at the meeting, which is a clear indication of the interest in and importance of the subject. This monograph includes chapters written by some of the presenters, and also a few other scientists who have made significant contribution in the field of asymmetric catalysis.

The symposium would not have been successful without invaluable support from the following academic, industrial, and professional organizations. The symposium was hosted and sponsored by the ACS Division of Organic Chemistry. Additional financial support for the symposium was received from the New Jersey Institute of Technology, the ACS Petroleum Research Fund, Merck Research Laboratories, National Starch and Chemical Company, DSM Fine Chemicals, Inc., Synthon Chiragenics Corporation, Strem Chemicals, Callery Chemical, Engelhard Corporation, Sorbent Technologies, and Aldrich Chemical Company. I am indebted to the presenters at the symposium, to the authors who contributed to this book, and to these organizations for their foresight in sponsoring this symposium and I am looking forward to continuing to work with them.

Sanjay V. Malhotra
Department of Chemistry and Environmental Science
New Jersey Institute of Technology
University Heights
Newark, NJ 07102–1982

Methodologies in Asymmetric Catalysis

Chapter 1

Chiral Dirhodium(II) Carboxamidates for Catalytic Asymmetric Synthesis

Michael P. Doyle

Department of Chemistry, University of Arizona, Tucson, AZ 85721

Chiral dirhodium(II) carboxamidates are remarkably effective catalysts for a variety of asymmetric transformations, especially those of diazoacetates and diazoacetamides, but also including Lewis acid catalyzed reactions. High selectivities are achieved even with low catalyst loadings, and polymer-bound catalysts are effective for multiple recovery and reuse. Cyclopropanation, carbon-hydrogen insertion, and ylide reactions have received considerable attention, and high enantiocontrol has been achieved, especially in intramolecular reactions.

The Catalysts

Dirhodium(II) carboxamidate catalysts are constructed from dirhodium(II) acetate by ligand replacement (eq 1) (*1-3*). Since the amide ligands are more

$$4LH + Rh_2(OAc)_4 \xrightarrow[\text{reflux}]{\text{PhCl}} Rh_2L_4 + 4HOAc \text{ (trapped with } Na_2CO_3) \quad (1)$$

susceptible to replacement in an acidic medium, the removal of acetic acid is the key element in the success of this replacement reaction. This is accomplished by trapping acetic acid as it forms (Soxhelet extractor containing sodium carbonate).

Four classes of dirhodium(II) carboxamides have been synthesized, and they are illustrated in Scheme 1: MEPY = methyl 2-oxopyrrolidine-carboxylate (*4*), MEOX = methyl 2-oxooxazolidine-carboxylate (*5*), MPPIM = methyl 3-phenylpropanoyl-2-oxoimidazolidine-carboxylate (*6*), and IBAZ = isobutyl 2-oxoazetidine-carboxylate (*7*). These catalysts are constructed with four bridging ligands so that two nitrogens and two oxygens reside on each rhodium, and the nitrogens (or oxygens) are cis to each other. This places the carboxylate attachment in two adjacent quadrants and allows convenient access to the substrate bound to the axial site of rhodium.

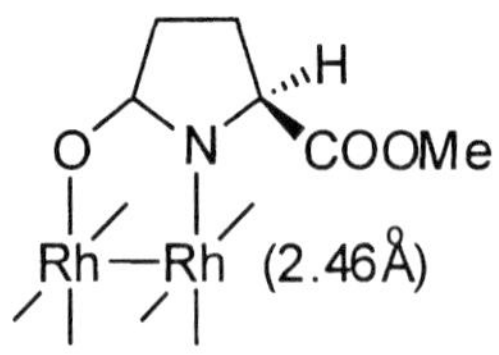

$Rh_2(5S\text{-}MEPY)_4$ (2.46Å) $Rh_2(4S\text{-}MEOX)_4$ (2.46Å)

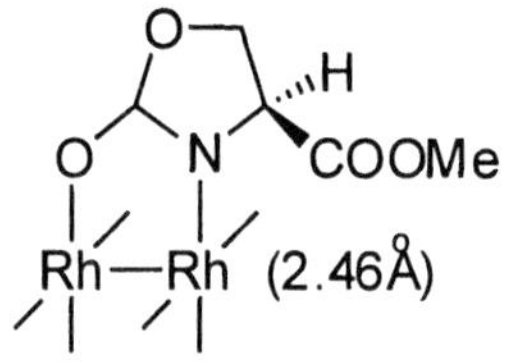

$Rh_2(4S\text{-}MPPIM)_4$ (2.46Å) $Rh_2(4S\text{-}IBAZ)_4$ (2.53Å)

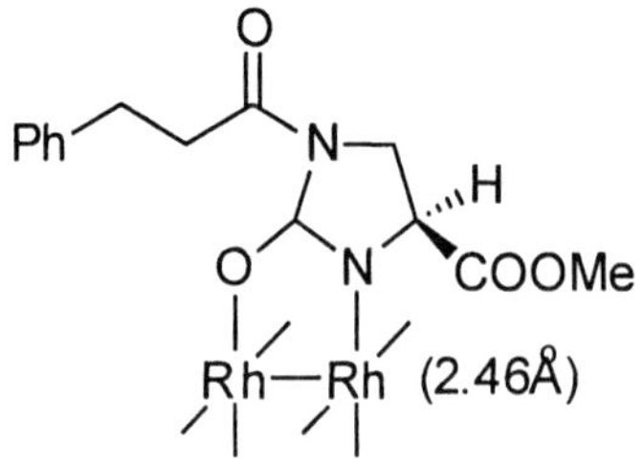

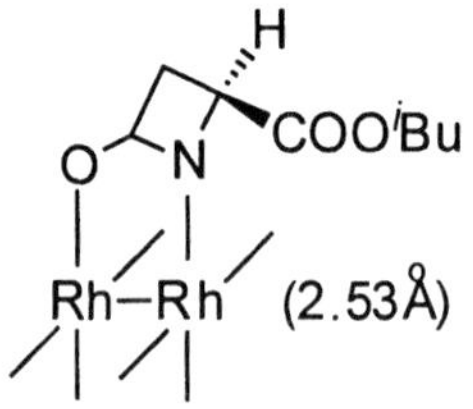

Scheme 1

Their preparation commonly results in the crystallization of the bis-acetonitrile complex in which acetonitriles occupy the axial coordination sites. These materials are then resistant to hydrolysis and to air oxidation, and they have been stored intact for several years.

The reactivities of these catalysts are less then those of dirhodium(II) carboxylates, but among the classes of carboxamidates in Scheme 1 there is significant variation. Azetidinone-ligated dirhodium catalysts like $Rh_2(4S\text{-}IBAZ)_4$ are the most reactive in an electrophilic sense. This is because the ligand NCO angle is larger, and its bite on dirhodium(II) lengthens the rhodium-rhodium bond (*7*).

Metal Carbene Chemistry

The uses of dirhodium(II) catalysts for asymmetric reactions that involve metal carbene intermediates are diverse and extensive (*2,8-10*). Some of the better known transformations are outlined in Scheme 2, which are applicable to an extensive listing of diazocarbonyl compounds. They include addition and insertion reactions, as well as association reactions leading to ylide intermediates.

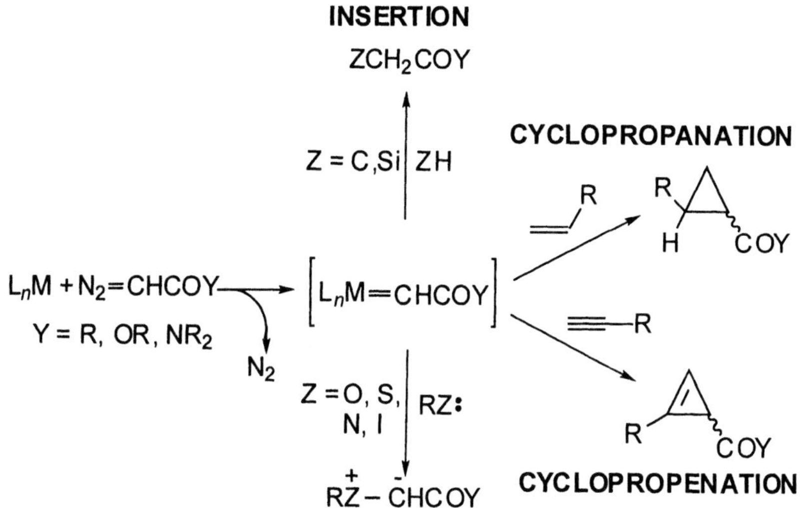

Scheme 2

Diazocarbonyl compounds are the reactive substrates for these reactions in which the rate limiting step is diazo decomposition in the formation of the metal carbene intermediate. Diazoacetates and diazoacetamides are prepared by standard methods from inexpensive reagents (*2*). They are handled conveniently without evidence of shock sensitivity, and they are stable to acids at p$H > 3$. The melting points of two diazoacetates suggest their stability:

Bu^t O Me–(C$_6$H$_2$)–O–C(O)–CHN$_2$ Bu^t

mp 152°C (*11*)

O O N–O–C(O)–CHN$_2$ O

mp 115°C (*12*)

Catalytic reactions take place through the intervention of a coordinatively unsaturated ligated metal complex, typically of copper(I) or rhodium(II), for which Lewis bases are inhibitory. Addition to a diazo compound produces an intermediate diazonium ion that, upon loss of dinitrogen, produces the metal carbene intermediate (Scheme 3). Transfer of the electrophilic carbene from the metal to the substrate forms the reaction product and returns the metal to its catalytically active state. The metal carbenes produced from copper and rhodium are highly electrophilic, and they behave more like metal-stabilized carbocations than as classical Fischer carbenes.

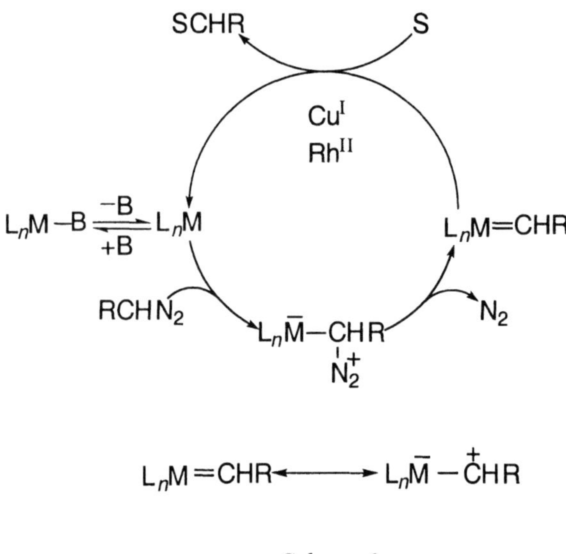

Scheme 3

Intramolecular Asymmetric Cyclopropanation

The use of chiral dirhodium(II) carboxamidates is especially advantageous for intramolecular cyclopropanation. As exemplified by the reaction with allyl diazoacetate (Scheme 4), use of the *S*-configured catalyst produces one product enantiomer with high % ee whereas use of the *R*-configured catalyst produces the other enantiomer with the same high % ee. Isolated yields (chromatography, distillation) are high in these reactions even when there is low catalyst loading (*13*).

A comparison of results for intramolecular cyclopropanation of allyl diazoacetates from the use of chiral rhodium, copper, and ruthenium catalysts (*14*) demonstrates the unique advantages of chiral dirhodium(II) carboxamidates (Scheme 5). Uniformly high enantiocontrol is achieved with the use of the

0.1 mol % Rh_2(5*R*-MEPY)$_4$, CH_2Cl_2, 72% — 95% ee

0.1 mol % Rh_2(5*S*-MEPY)$_4$, CH_2Cl_2, 75% — 95% ee

Rh_2(5*R*-MEPY)$_4$ Rh_2(5*S*-MEPY)$_4$

Scheme 4

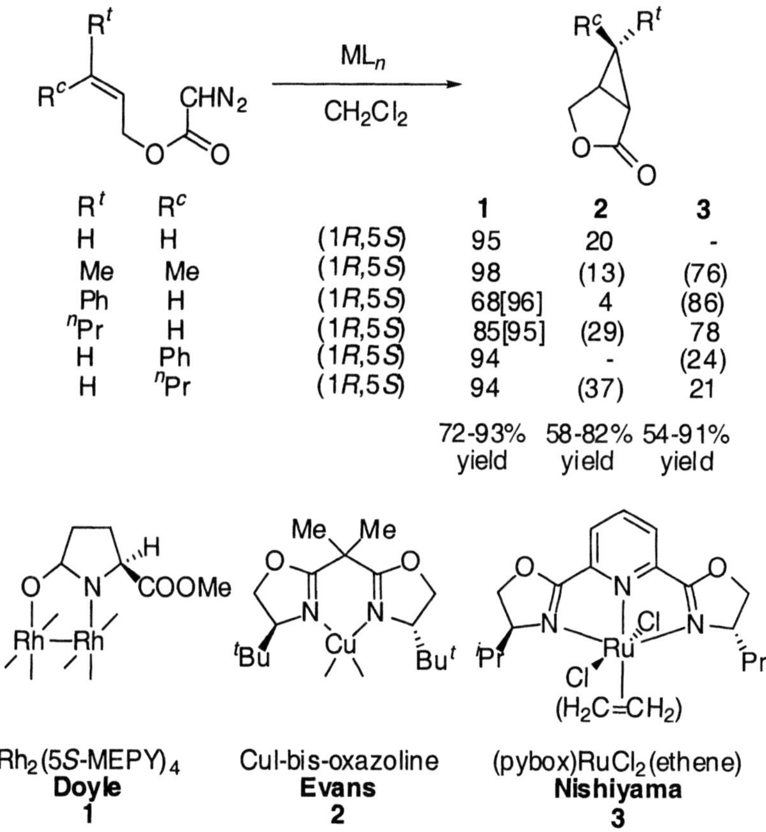

R^t	R^c		1	2	3
H	H	(1*R*,5*S*)	95	20	-
Me	Me	(1*R*,5*S*)	98	(13)	(76)
Ph	H	(1*R*,5*S*)	68[96]	4	(86)
^{n}Pr	H	(1*R*,5*S*)	85[95]	(29)	78
H	Ph	(1*R*,5*S*)	94	-	(24)
H	^{n}Pr	(1*R*,5*S*)	94	(37)	21
			72-93% yield	58-82% yield	54-91% yield

Scheme 5

$Rh_2(MEPY)_4$ catalysts, except for addition to the *trans*-distributed carbon-carbon double bonds in which case the $Rh_2(MPPIM)_4$ catalysts return enantioselectivities to ≥95% ee. Similarly high enantioselectivities are obtained with allylic diazoacetamides (*15*), and somewhat lower selectivities are reported for homoallylic systems (*13*, *15*).

This selectivity has recently been extended to phenyldiazoacetates that are typically unreactive with dirhodium(II) carboxamidates (*16*). The azetidinone-ligated dirhodium(II) catalysts, being more reactive towards diazo decomposition, are effective for these transformations (*17*). Application to the synthesis of milnacipran and some of its analogues (Scheme 6) is illustrative (*18*). Here, turnover numbers up to 10,000 were effective, and high enantioselectivities could be achieved.

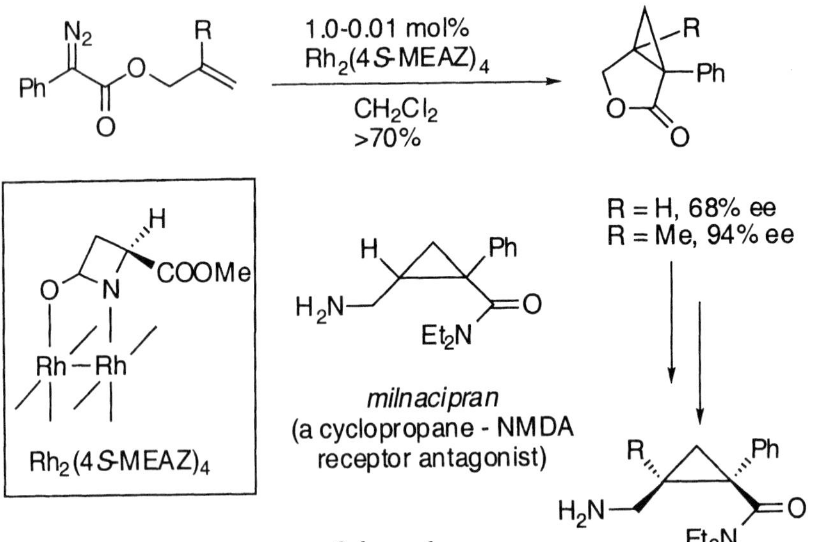

1.0-0.01 mol% $Rh_2(4S\text{-}MEAZ)_4$

CH_2Cl_2

>70%

R = H, 68% ee
R = Me, 94% ee

$Rh_2(4S\text{-}MEAZ)_4$

milnacipran
(a cyclopropane - NMDA receptor antagonist)

Scheme 6

An extension of these catalysts for *cis*-selective intermolecular cyclopropanation has been achieved. We had previously observed that $Rh_2(4S\text{-}IBAZ)_4$ gave the cis isomer predominantly and with high % ee (*7*). Subsequently, Katsuki and coworkers had developed their salen-bino ligand (Scheme 7) for exceptional selectivity in reactions of *tert*-butyl diazoacetate with styrene (*19,20*). We extended our earlier work by using a chiral alkyl group on the ester attachment for the azetidinone ligand, Rh_2(*S,R*-MenthAZ)$_4$, and this improved diasteroselectivity for reactions with styrene (*21*). Application to the synthesis of a urea-PETT analog (Scheme 8), demonstrated even higher stereoselectivity. With 2,4,6-trimethylstyrene diastereocontrol was 92:8 (cis:trans), and the cis isomer was produced in 97% ee (*21*).

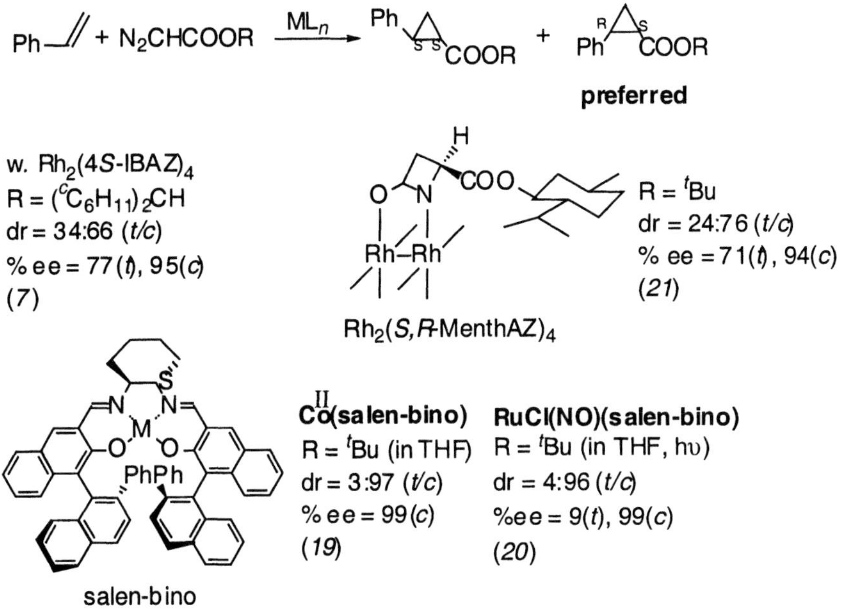

Ph + N_2CHCOOR —ML_n→ Ph (S S) COOR + Ph (R S) COOR

preferred

w. Rh_2(4*S*-IBAZ)$_4$
R = ($^c C_6H_{11}$)$_2$CH
dr = 34:66 (*t/c*)
% ee = 77 (*t*), 95(*c*)
(*7*)

Rh_2(*S,R*-MenthAZ)$_4$
R = ^{t}Bu
dr = 24:76 (*t/c*)
% ee = 71(*t*), 94(*c*)
(*21*)

salen-bino

Co^{II}(salen-bino)
R = ^{t}Bu (in THF)
dr = 3:97 (*t/c*)
% ee = 99 (*c*)
(*19*)

RuCl(NO)(salen-bino)
R = ^{t}Bu (in THF, hυ)
dr = 4:96 (*t/c*)
%ee = 9(*t*), 99(*c*)
(*20*)

Scheme 7

Intramolecular Asymmetric Carbon-Hydrogen Insertion

One of the most useful and potentially valuable transformations of diazoacetamides is insertion into a C-H bond to form a five-membered ring product. With diazoacetates these reactions occur to form γ-lactone products, and using the Rh_2(MPPIM)$_4$ catalysts this process occurs with exceptional selectivity (Scheme 9) (*2*). With Rh_2(4*S*-MPPIM)$_4$ the *S*-enantiomer of the γ-lactone product is formed, whereas with Rh_2(4*R*-MPPIM)$_4$ the *R*-enantiomer of

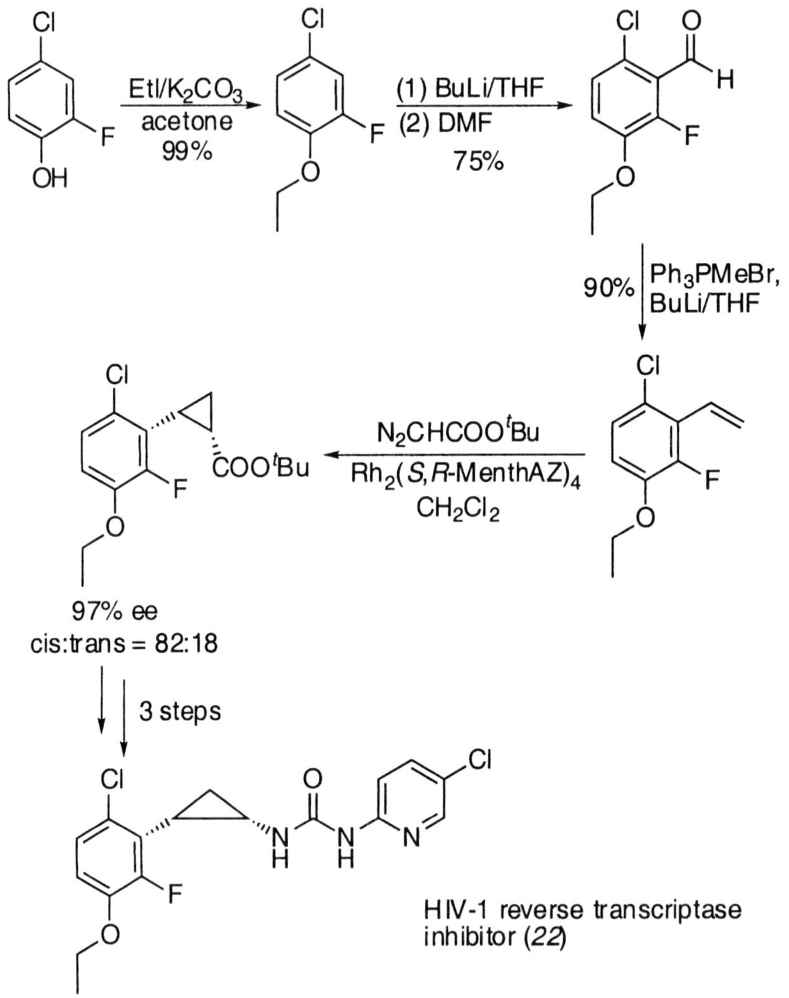

HIV-1 reverse transcriptase inhibitor (*22*)

Scheme 8

the γ-lactone product is preferentially obtained (*23*). Yields are good, and regioselectivity is > 20:1. This methodology has been applied to the synthesis of lignans (Scheme 10) that are readily accessible from 3-substituted-butyrolactones (Scheme 10) (*23*). Applications to the total syntheses of *R*-baclofen (*24*), deoxyxylolactone (Scheme 11) (*25,26*), and deoxyxylolactam (*27*) have been reported.

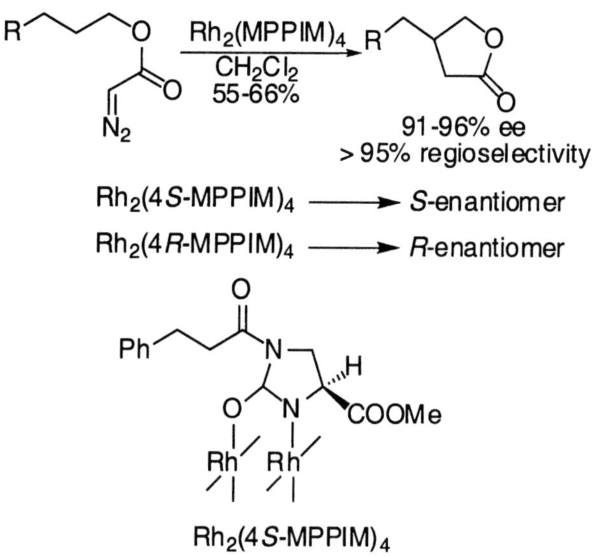

$Rh_2(MPPIM)_4$, CH_2Cl_2, 55-66%

91-96% ee
> 95% regioselectivity

$Rh_2(4S\text{-MPPIM})_4$ → *S*-enantiomer
$Rh_2(4R\text{-MPPIM})_4$ → *R*-enantiomer

$Rh_2(4S\text{-MPPIM})_4$

Scheme 9

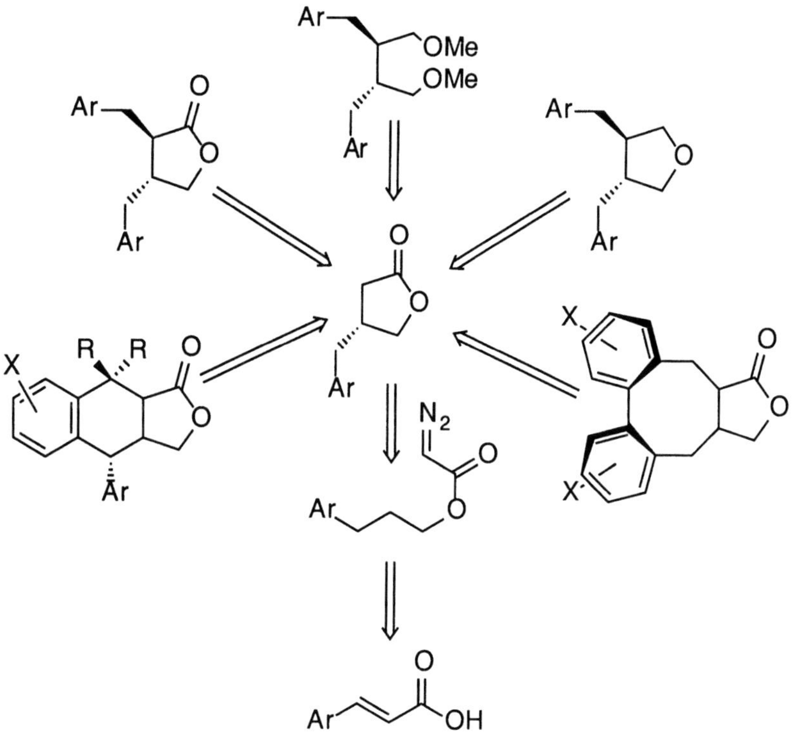

Scheme 10

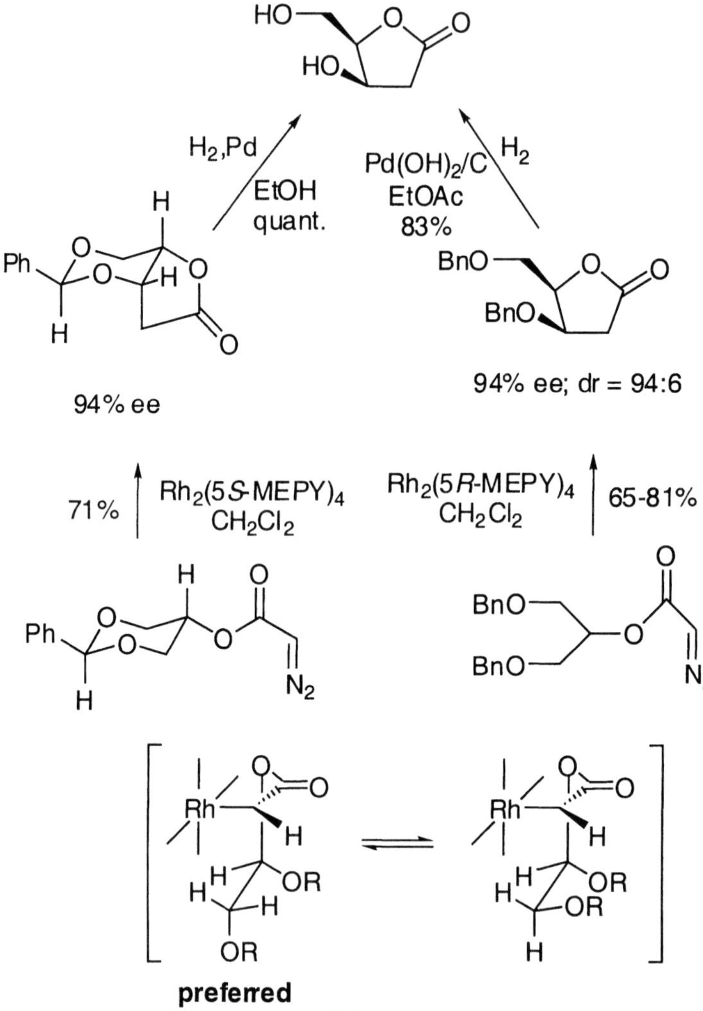

Scheme 11

In defining the preferential selectivities in the formation of deoxyxylolactone, diastereocontrol is obviously determined by the preference of the OR group (Scheme 11) for an orientation anti to the ester's ether oxygen. These same or similar control features can be seen in reactions of cyclohexyl diazoacetates where preferential equatorial C-H insertion occurs because the metal carbene unit is in the axial conformation (*28*). In these cases it is the face of the catalyst surface with its attached substituents that places a steric barrier on the reactive conformation of the intermediate metal carbene.

Another facet of these transformations is the ability of chiral substrates to distinguish match and mismatch with the chiral dirhodium(II) carboxamidate (*28, 29*). This feature is clearly evident in reactions of steroidal diazoacetates (for example, Scheme 12) (*30*). Here the *S*-configured cholesterol undergoes preferential insertion into the 2-position with *R*-configured dirhodium(II) carboxamidates, but insertion into the 3-position to form a strained β-lactone occurs with the *S*-configured catalyst. Access to the 4-position is restricted.

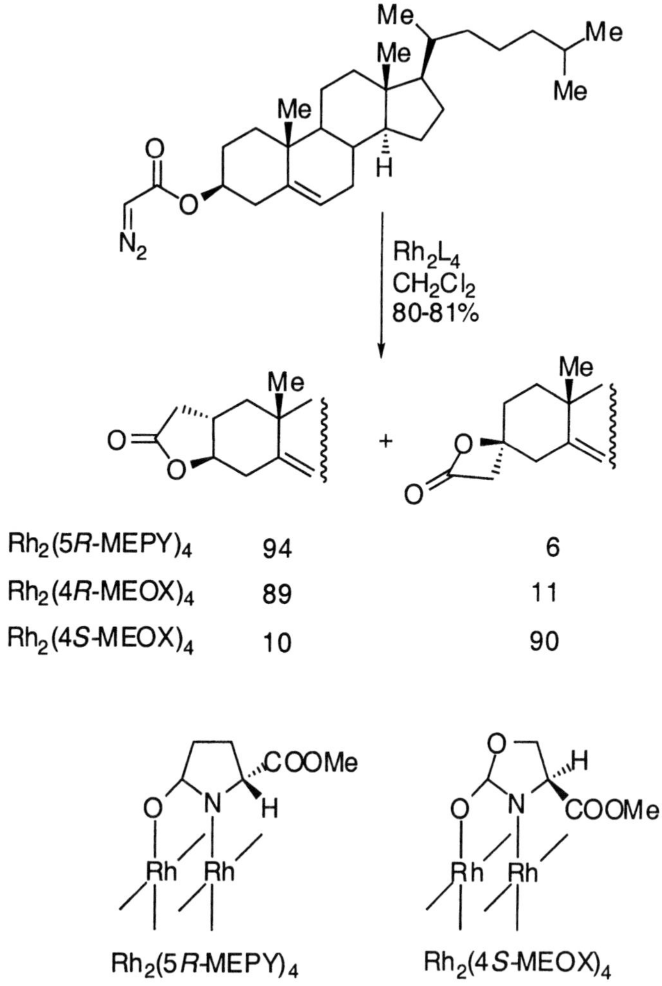

Rh_2L_4, CH_2Cl_2, 80-81%

Rh_2(5*R*-MEPY)$_4$	94	6
Rh_2(4*R*-MEOX)$_4$	89	11
Rh_2(4*S*-MEOX)$_4$	10	90

Rh_2(5*R*-MEPY)$_4$ Rh_2(4*S*-MEOX)$_4$

Scheme 12

Rhodium(II) Catalysts as Chiral Lewis Acids

Evidence has recently been reported of the catalytic use of chiral dirhodium(II) carboxamidates for the hetero-Diels-Alder reaction (eq 2) (*31*).

O Ar + OTMS OMe $\xrightarrow[CH_2Cl_2]{Rh_2L_4}$ $\xrightarrow{TFA}$ Ar O O (2)

up to 95% ee

Here the major advantage of these catalysts is that they can be employed with TON of 10,000, which is a vast improvement over applications of other chiral Lewis acids to the same transformation.

Literature Cited

1. Doyle, M. P.; Ren, T. *Prog. Inorg. Chem.* **2001**, *49*, 113-168.
2. Doyle, M. P.; McKervey, M. A.; Ye, T. *Modern Catalytic Methods for Organic Synthesis with Diazo Compounds*; John Wiley & Sons, Inc.: New York, NY, 1998.
3. Doyle, M. P. in *Catalysis by Di- and Polynuclear Metal Cluster Complexes*; Adams, R. D.; Cotton, F. A., Eds.; VCH Publishers: New York, NY, 1998; chapter 7.
4. (a) Doyle, M. P.; Winchester, W. R.; Hoorn, J. A. A.; Lynch, V.; Simonsen, S. H.; Ghosh, R. *J. Am. Chem. Soc.* **1993**, *115*, 9968-9978. (b) Doyle, M. P.; Winchester, W. R.; Protopopova, M. N.; Kazala, A. P.; Westrum, L. *J. Org. Syn.* **1996**, *73*, 13-24.
5. Doyle, M. P.; Dyatkin, A. B.; Protopopova, M. N.; Yang, C. I.; Miertschin, C. S.; Winchester, W. R.; Simonsen, S. H.; Lynch, V.; Ghosh, R. *Recl. Trav. Chim. Pays-Bas* **1995**, *114*, 163-170.
6. Doyle, M. P.; Zhou, Q.-L.; Raab, C. E.; Roos, G. H. P.; Simonsen, S. H.; Lynch, V. *Inorg. Chem.* **1996**, *35*, 6064-6073.
7. Doyle, M. P.; Zhou, Q.-L.; Simonsen, S. H.; Lynch, V. *Synlett* **1996**, 697-698.
8. Doyle, M. P. In *Catalytic Asymmetric Synthesis*, Second Edition; Ojima, I., Ed.; John Wiley & Sons, Inc.: New York, NY, 2000; chapter 5.
9. Doyle, M. P.; Forbes, D. C.; *Chem. Rev.* **1998**, *98*, 911-935.
10. Davies, H. M. L.; Antoulinakis, E. G. *Org. Rxns.* **2001**, *57*, 1-326.
11. Doyle, M. P.; Bagheri, V.; Wandless, T. J.; Harn, N. K.; Brinker, D. A.; Eagle, C. T.; Loh, K.-L. *J. Am .Chem. Soc.* **1990**, *112*, 1906-1912.
12. Doyle, M. P.; Kalinin, A. V. *J. Org. Chem.* **1996**, *61*, 2179-2184.

13. Doyle, M. P.; Austin, R. E.; Bailey, A. S.; Dwyer, M. P.; Dyatkin, A. B.; Kalinin, A. V.; Kwan, M. M. Y.; Liras, S.; Oalmann, C. J.; Pieters, R. J.; Protopopova, M. N.; Raab, C. E.; Roos, G. H. P.; Zhou, Q.-L.; Martin, S. F. *J. Am. Chem. Soc.* **1995**, *117*, 5763-5775.
14. Doyle, M. P.; Peterson, C. S.; Zhou, Q.-L.; Nishiyama, H. *J. Chem. Soc., Chem. Commun.* **1997**, 211-212.
15. Doyle, M. P.; Kalinin, A. V. *J. Org. Chem.* **1996**, *61*, 2179-2184.
16. Doyle, M. P.; Zhou, Q.-L.; Charnsangavej, C.; Longoria, M. A.; McKervey, M. A.; Garcia, C. F. *Tetrahedron Lett.* **1996**, *37*, 4129-4132.
17. Doyle, M. P.; Davies, S. B.; Hu, W. *Org. Lett.* **2000**, *2*, 1145-1147.
18. Doyle, M. P.; Hu, W. *Adv. Synth. & Catal.* **2001**, *343*, 299-302.
19. Niimi, T.; Uchida, T.; Irie, R.; Katsuki, T. *Tetrahedron Lett.* **2000**, *41*, 3647-3651.
20. Uchida, T.; Irie, R.; Katsuki, T. *Tetrahedron* **2000**, *56*, 3501-3509.
21. Hu, W.; Timmons, D. J.; Doyle, M. P. *Org. Lett.* **2002**, *4*, 901-904.
22. Högberg, M.; Sahlberg, C.; Engelhardt, P.; Noréen, R.; Kangasmetsä, J.; Johansson, N. G.; Oberg, B.; Vrang, L.; Zhang, H.; Salhberg, B. L.; Unge, T.; Lövgren, S.; Fridborg, K.; Bäckbro, K. *J. Med. Chem.* **1999**, *42*, 4150.
23. Bode, J. W.; Doyle, M. P.; Protopopova, M. N.; Zhou, Q.-L. *J. Org. Chem.* **1996**, *61*, 9146-9155.
24. Doyle, M. P.; Hu, W. *Chirality* **2002**, *14*, 169-172.
25. Doyle, M. P.; Dyatkin, A. B.; Roos, G. H. P.; Cañas, F.; Pierson, D. A.; van Basten, A.; Müller, P.; Polleux, P. *J. Am. Chem. Soc.* **1994**, *116*, 4507-4508.
26. Doyle, M. P.; Tedrow, J. S.; Dyatkin, A. B.; Spaans, C. J.; Ene, D. G. *J. Org. Chem.* **1999**, *64*, 8907-8915.
27. Doyle, M. P.; Phillips, I. M.; Timmons, D. J. *Adv. Synth. & Catal.* **2002**, *344*, 91-95.
28. Doyle, M. P.; Kalinin, A. V.; Ene, D. G. *J. Am. Chem. Soc.* **1996**, *118*, 8837-8846.
29. Doyle, M. P.; Kalinin, A. V. *Tetrahedron Lett.* **1996**, *37*, 1371-1374.
30. Doyle, M. P.; Davies, S. B.; May, E. J. *J. Org. Chem.* **2001**, *66*, 8112-8119.
31. Doyle, M. P.; Phillips, I. M.; Hu, W. *J. Am. Chem. Soc.* **2001**, *123*, 5366-5367.

Chapter 2

Anchored Homogeneous Catalysts: The Best of Both Worlds

R. Augustine, S. Tanielyan, S. Anderson, Y. Gao, P. Goel, N. Mahata, J. Nair, C. Reyes, H. Yang, and A. Zsigmond

Center for Applied Catalysis, Department of Chemistry and Biochemistry, Seton Hall University, South Orange, NJ 07079

Homogeneous catalysts can be anchored to a number of different support materials using heteropoly acids as the anchoring agents. This technique can be used to anchor pre-formed complexes thus eliminating the need to modify the ligand to accomplish the 'heterogenization'. These anchored catalysts are stable in a range of solvents with little, if any, loss of the complex during the reaction. They are generally as active and selective as the homogeneous analogs and are particularly effective with the chiral complexes commonly employed today for enantioselective hydrogenations. These catalysts have been used for multiple batch and continuous hydrogenations with TON's of 10,000 and greater and TOF's as high as 650 hr^{-1}.

The concept of "heterogenizing" homogeneous catalysts was first introduced over thirty years ago (*1,2*). It was considered that by converting a homogeneous catalyst into a heterogeneous species one would be able to combine the activity and selectivity of the homogeneous catalyst with the ease of separation inherent to a heterogeneous catalyst and, thus, have the 'best of both worlds'. The earliest reports of work in this area involved the use of a solid, heterogeneous, ligand to which the metal was then attached.

This approach is still the most commonly used (*3,4*). However, there are several problems associated with these systems. One of the more important is concerned with the preparation of the solid ligand. While this approach is relatively straightforward with simple ligands such as triphenyl phosphine, converting the more complex, enantioselective ligands commonly used today into some heterogeneous form can require considerable synthetic prowess. Further, these so-called 'tethered' catalysts are prone to loss of metal during the reaction, are frequently not as active or selective as the homogeneous analog, and usually lose their activity on attempted re-use (*5-7*).

What is needed is a method of anchoring preformed complexes to a support material to give species which will retain the activity and selectivity of the homogeneous catalyst, even on re-use. Direct attachment of the typical complex to a support material is not generally observed, but it has recently been reported that some complexes have been attached to MCM-41 giving active, stable catalysts (*8*). However, this appears to be the only material capable of such direct interaction.

We describe here our approach to this problem in which we have anchored a number of catalytically active homogeneous complexes to a variety of support materials using heteropoly acids as the anchoring agents. The resulting catalysts are generally at least as active and selective as the homogeneous counterpart, are easily separated from the reaction mixture, can be readily adapted for use in continuous reactors, and do not show any significant metal loss during either multiple batch reactions or those run in a continuous system (*9-13*).

These anchored homogeneous catalysts are prepared in a two step procedure. First, a solution of the heteropoly acid (20 μmoles in 2.5 mL of alcohol) is added with vigorous stirring to a suspension of the support material (300mg in 10 mL of alcohol) with stirring continued for about three hours followed by the removal of the liquid and thorough washing of the solid. The resulting solid is then suspended in another 10mL of degassed alcohol and a solution of the homogeneous catalyst (20 μmoles in 1 mL of alcohol) is added under an inert atmosphere, with stirring, over a 30 min period. Stirring is continued for 8 to 12 hours under an inert atmosphere, the liquid removed and the solid washed thoroughly until no color is observed in the wash liquid. This material can be used directly or dried for future use. Typical loading is about 7 – 10% with respect to the complex which corresponds to about 0.4 – 0.5% metal.

A description of the nature of these catalysts has been published previously (*13,14*). The important factor with these catalysts is that there is an interaction between the metal atom of the complex and the heteropoly acid so that the ligand is not involved in the anchoring process. This permits the heterogenization of pre-formed complexes, a much simpler approach than one requiring the preparation of a heterogeneous ligand from which to make the complex. Since the ligand is not involved in the anchoring process, there is virtually no significant limitation on the types of ligands which may be used. Further, this anchoring procedure can be used with a number of different support materials in a variety of solvents. These catalysts have also been shown to be stable in extensive re-use and continuous applications with very little, if any, metal loss in the process.

We had, however, frequently seen an increase in activity and selectivity when these anchored homogeneous catalysts were re-used. Since the first use of these catalysts involved exposure to hydrogen during the reaction, it was thought that this increase in activity was caused by some hydrogen induced modification of the catalyst. As part of our previous work we had found that the catalyst activity and selectivity could be enhanced by pre-hydrogenating the anchored complex for several hours before introduction of the substrate. A detailed description of this problem will be published elsewhere (*14*).

Reaction Parameters

Over the past several years a considerable amount of work has been done on examining the effect which the various reaction parameters have on the stability, activity and selectivity of these anchored homogeneous catalysts. Most of the hydrogenations were run at temperatures ranging from room temperature to about 50° - 60°C and pressures from atmospheric to about 100 psig. All reactions were run at constant volume – constant pressure using computerized monitoring of the hydrogen uptake as well as reaction pressure and temperature (*15*).

Ligands

A number of chiral rhodium complexes anchored on alumina by PTA were used for the enantioselective hydrogenation of methyl 2-acetamidoacrylate (**1**). The data in Table 1 lists the rates and ee data for these reactions as well as the corresponding data obtained using the homogeneous catalyst. The anchored catalysts were not pre-hydrogenated so data are presented for the first and third use of these species.

$$H_2C{=}C(NAc)CO_2Me \ (\mathbf{1}) \xrightarrow[\text{Rh (Ligand)/ PTA / }Al_2O_3]{H_2} H_3C\text{-}CH(NAc)\text{-}CO_2Me\ (R) + H_3C\text{-}CH(NAc)\text{-}CO_2Me\ (S)$$

Table 1. Rate data for the hydrogenation of 1 over Rh(ligand)/PTA/Al_2O_3 and the homogeneous analog.

		Anchored		*Homogeneous*	
Ligand	*Use #*	*Rate*[a]	*ee%*	*Rate*[a]	*ee%*
DiPamp	1	0.32	90	0.25	76
	3	1.67	92		
Prophos	1	2.0	68	0.26	66
	3	2.6	63		
Me-DuPhos	1	1.8	83	3.3	96
	3	4.4	95		
BPPM	1	3.8	21	7.4	84
	3	8.2	87		
Josiphos	1	2.5	44	8.9	75
	3	8.7	87		

a mmole H_2 / mmole Rh / min.

Other chiral ligands, such as Binap and tetraME-BITIOP (*16*), were also used in anchored complexes involved in the enantioselective hydrogenation of various prochiral substrates as described in the following sections. The achiral ligands Triphenyl phosphine, dppe, dppb, DiPFc (*17*), bis dipentaflourophenyl-phosphinoethane and bis diphenylphosphinoferrocene, have been used to prepare catalysts for the hydrogenation of substrates such as 1-hexene and carvone.

Metal Complexes

Almost all of the work done to date has involved the anchoring of rhodium complexes. There has been an assumption made that the only types of complexes which can be used in our anchoring procedure are cationic species (*17*). We have, however, successfully anchored Wilkinson's catalyst, a non-cationic complex. We have also prepared anchored tris(triphenylphosphine)-rhodium catalysts starting from the tris(triphenylphosphine)rhodium chloride and bromide, and treating the anchored $Rh(COD)_2$ precursor with triphenyl phosphine. These catalysts, as well as the commercially available phosphinated polystyrene supported Wilkinsons (obtained from Strem Chemicals) were used for the hydrogenation of 1-hexene in ethanol at room temperature and atmospheric pressure. The catalysts prepared from the chloride or bromide as well as from the anchored $Rh(COD)_2$ precursor were all comparable in activity. The polymer anchored Wilkinson was about an order of magnitude less active.

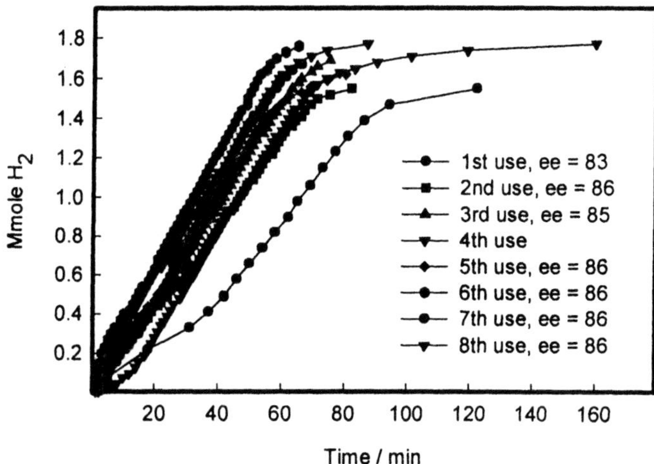

Figure 1. Hydrogen uptake curves for the hydrogenation of 2 over Ru(tetraMe-BITIOP)/PTA/Al_2O_3 at 50°C and 55 psig.

We have also done some work using anchored ruthenium complexes. Figure 1 shows the hydrogen uptake data obtained on hydrogenation of dimethyl itaconate (**2**) over Ru(tetraMe-BITIOP)/PTA/Al_2O_3 at 50°C and 55 psig. Analysis of the combined reaction mixtures showed that ruthenium was present in less than 0.5 ppm.

H_2C CO_2Me / CO_2Me **2** — Ru(Binap)/**HPA**/Al_2O_3, 50°C 55 psig → H_3C CO_2Me / CO_2Me (R) + H_3C CO_2Me / CO_2Me (S)

Heteropoly Acids

Some preliminary work, listed in Table 2, was done which showed that heteropoly acids (HPA's) other than phosphotungstic acid (PTA) could be used as anchoring agents for homogeneous catalysts. The data shown in Table 3 for the hydrogenation of **1** over anchored Rh(DiPamp) catalysts is interesting in that contrary to the decrease in ee with increasing pressure that is commonly observed with this catalyst and substrate, this is not found with reactions run using PMA as the anchoring agent. Table 4 shows the reaction data for the HPA effect on the hydrogenation of **2** over anchored Ru(Binap) catalysts.

Table 2. Heteropoly acid effect on the hydrogenation of 1 over Rh(DiPamp)/HPA/Al_2O_3

		Non-Prehydrogenated		*Prehydrogenated*	
Heteropoly Acid	*Use #*	*Rate[a]*	*ee%*	*Rate[a]*	*ee%*
Phosphotungstic	1	0.25	84	0.92	80
	3	0.92	85		
Silicotungstic	1	0.21	79	1.14	81
	3	0.96	80		
Phosphomolybdic	1	0.36	86	1.28	87
	3	0.94	86		
Silicomolybdic	1	0.18	84	0.67	85
	3	0.75	85		

a mmole H_2 / mmole Rh / min.

Table 3. Effect of the heteropoly acid and pressure on the hydrogenation of 1 over Rh(DiPamp)/HPA/Al_2O_3

		15 psig		*30 psig*	
Heteropoly acid	*Use #*	*Rate[a]*	*ee%*	*Rate[a]*	*ee%*
Phosphotungstic	1	0.019	80	0.026	64
	4	0.008	79	0.020	64
Silicotungstic	1	0.022	85	0.034	69
	4	0.009	81	0.023	69
Phosphomolybdic	1	0.028	90	0.035	89
	4	0.024	91	0.027	90
Silicomolybdic	1	0.009	86	0.027	69
	4	0.007	83	0.016	70

a mmole H_2 / min.

Table 4. Heteropoly acid effect on the hydrogenation of 2 over Ru(Binap)/HPA/Al_2O_3

Heteropoly Acid	*Use #*	*Rate*[a]	*ee%*
Phosphotungstic	1	0.028	85
	5	0.034	72
	11	0.035	68
Silicotungstic	1	0.052	84
	4	0.036	85
	7	0.032	83
Silicomolybdic	1	0.047	86
	3	0.067	91
	6	0.048	94

a mmole H_2 / min.

Supports

We have been successful in anchoring homogeneous catalysts to a number of different supports. The data in Table 5 illustrate some of our earliest work in this area. Because of the apparent better reaction rate observed with the alumina supported catalyst, this material was used as the primary support for most of the work done to date. It was recognized, however, that there are different types of alumina , but after examining alumina samples from a number of sources the material selected for use was the neutral gamma alumina available from Strem Chemicals.

Table 5. Support effect in the hydrogenation of 1 over Rh(DiPamp)/PTA/support

Support	*Use #*	*Rate*[a]	*ee%*
Montmorillonite K	1	0.18	76
	3	0.56	91
Carbon	1	0.07	83
	3	0.4	90
Alumina	1	0.32	90
	3	1.67	95
Lanthana	1	0.38	91
	3	0.44	92

a mmole H_2 / mmole Rh / min.

CO_2H / NAc — $\xrightarrow[\text{1 Atm } H_2 \text{ 50°C}]{Rh(BPPM)/PTA/Al_2O_3}$ — H CO_2H NAc (R) + H CO_2H NAc (S)

3

Samples of this alumina were washed with ethanol to remove the fines and then were used for catalyst preparation as described above. Figure 2 shows the hydrogen uptake curves for successive hydrogenations of 2-acetamidocinnamic acid (**3**) over an alumina anchored Rh(BPPM) catalyst.

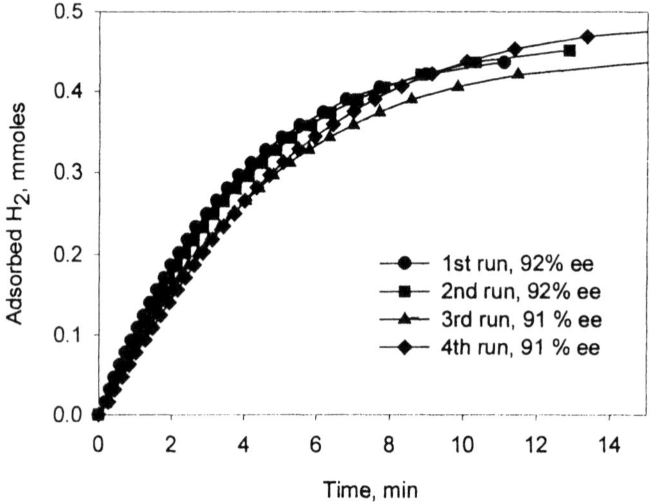

Figure 2. Hydrogen uptake curves for the hydrogenation of ***3*** *over Rh(BPPM)/PTA/Al_2O_3 at 50°C and atmospheric pressure.*

Several different carbons were also examined as potential supports. Some early work used Norit and Darco, but we have since settled on C-28 carbon from Johnson Matthey as our support material. The standard procedure described above for the preparation of the anchored alumina catalysts was also followed using a carbon support.

Figures 3a and b show the uptake curves observed for the hydrogenation of **3** over Rh(DuPhos)/PTA/Al_2O_3 and Rh(DuPhos)/PTA/C, respectively. Each catalyst contains about the same amount of complex with the carbon supported material more active and selective than the alumina anchored catalyst.

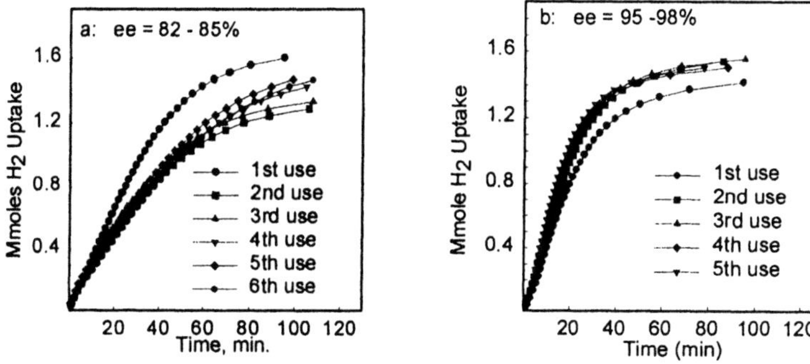

Figure 3. Hydrogen uptake data for the hydrogenation of 3 at 50°C and atmospheric pressure over anchored Rh(DuPhos) catalysts: a, alumina support; b, carbon support.

Solvents

Some questions have been raised concerning the stability of these anchored catalysts in solvents other than absolute ethanol, the material most commonly been used in all of the preliminary work (*17*). To examine this problem, we ran a number of hydrogenations with different anchored complexes in various solvents to determine the effect of the solvent on both catalyst stability and activity.

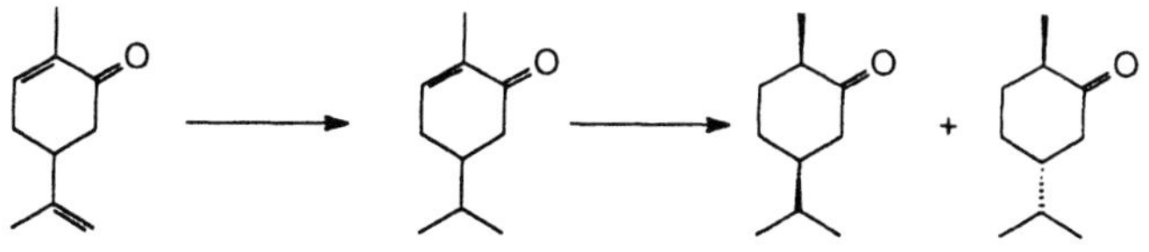

Table 6 lists the effect of solvent on the TOF's in the hydrogenation of carvone (**4**) over the homogeneous Wilkinson's catalyst (**6**), Wilkinson/PTA/Al_2O_3 (**7**), Wilkinson/STA/Al_2O_3 (**8**),and polystyrene supported Wilkinson (**9**) catalysts. It can be seen that the solvent has virtually no effect on those reactions run using the homogeneous catalyst but has a significant effect on the hydrogenations run using the heteropoly acid anchored species. The commercially available phosphinated polystyrene supported material, **9**, is relatively inactive with the solvent used having only a slight effect on the rate of reaction with this catalyst.

Table 6. Solvent effect on the TOF (min^{-1}) in the hydrogenation of 4 over homogeneous and supported Wilkinson's catalysts

Catalyst	*Abs EtOH*	*i-PrOH*	*EtOH-2B*[a]	*95% EtOH*	*EtOAc*	*ϕ-CH_3*	*THF*
6	1.82	1.67	1.25	1.05	1.74	1.70	
7	4.70	2.90	1.25	1.30	0.77	0.01	0.05
8	3.96	2.14	1.50	0.44	0.33		0.30
9	0.29	0.34	0.26	0.24	0.50	0.60	0.30

a EtOH denatured with 0.5% i-PrOH.

These reactions were stopped after one equivalent of hydrogen was taken up so they represent the rate of formation of the dihydro-carvone (**5**). The solid catalysts were used several times and the rates are the average of these reactions. Also measured was the effect of solvent on the amount of rhodium lost in each of those reactions run using a solid catalyst. Table 7 lists the ppm of rhodium found in the first reaction mixture for each solvent and catalyst. Thus, this number represents the maximum amount of rhodium in these reaction mixtures. It can be seen that the maximum loss was found with the solvents in which the reaction was the slowest. Further, the polystyrene supported material shows the most significant rhodium loss.

Table 7. Solvent effect on the leaching of Rh[a] in the hydrogenation of 4 over supported Wilkinson's catalysts

Catalyst	*Abs. EtOH*	*i-PrOH*	*EtOH-2B*[b]	*95% EtOH*	*EtOAc*	*ϕ-CH_3*	*THF*
7	0.28	0.18	0.21	0.74	1.16	2.5	3.1
8	0.10	0.07	0.05	0.24	0.19		1.2
9	3.2	2.0	4.5	4.5	1.6	5.6	38.0

a ppm of Rh in first reaction mixture.

b EtOH denatured with 0.5% i-PrOH.

Large Scale and Continuous Reactions

Most of the reactions discussed so far have used small amounts of substrate in order to acquire a large amount of reaction data in a reasonable length of time. For the most part, these reactions used substrate/catalyst ratios of 50 to 100. Since these low ratios are not indicative of any potential commercial use of these catalysts, several anchored homogeneous catalysts were used in either large scale multiple batch reactions or in a continuous reaction using a continuous stirred tank reactor (CSTR).

The hydrogenation of several successive 5 mL portions of 1-hexene over an anchored Rh(dppb) catalyst proceeded with no loss of activity between runs. The overall TON in this reaction was about 10,000 and analysis of the combined reaction mixtures showed that rhodium was present in less than 1ppm. The hydrogen uptake curves for another example of a large scale multiple batch reaction are shown in Fig. 4. Here an anchored Rh(Me-DuPhos) was used to hydrogenate 10mL portions of **2** with an overall TON of 10,000. The TOF for the reaction was 660 hr^{-1}. and the product ee was 97% - 98%. No rhodium was detected in the product.

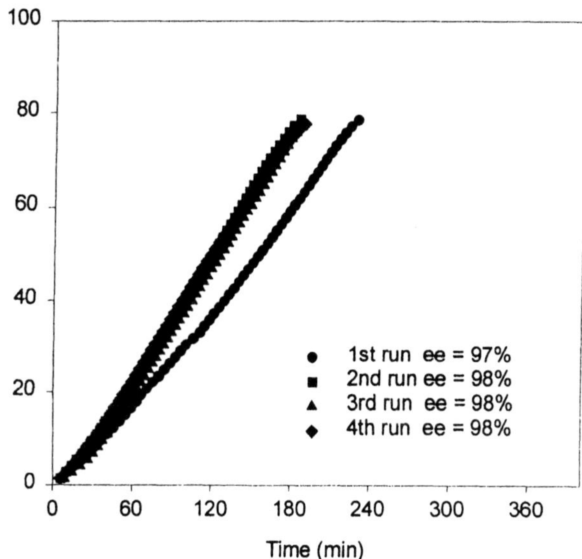

Figure 4. Hydrogen uptake data for the hydrogenation of 10 mL portions of 2 over Rh(DuPhos)/PTA/Al_2O_3 at R.T. and 50 psig.

A continuous hydrogenation of 1-hexene over anchored Rh(dppb) was run for 28 hours with a TON of 17,000 and no break in the rate of hydrogen uptake. Analysis of the product showed that rhodium was present in less than 0.2ppm.

Figure 5 shows a detailed analysis of a continuous hydrogenation of **2** over anchored Rh(Me-DuPhos) at room temperature and 50 psig. The slight break in the data curve at about 5 days was the result of a change in the substrate stream. The original material was used up and replaced by some obtained from another source. No complex was detected in the reaction mixture.

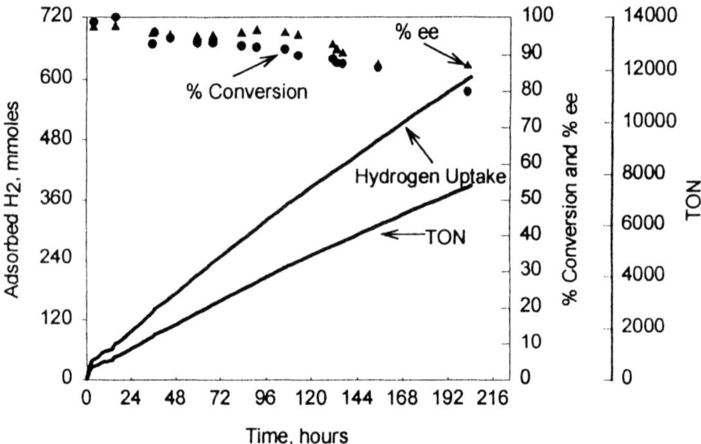

Figure 5. Reaction data for the CSTR hydrogenation of 2 over Rh(DuPhos)/PTA/Al_2O_3 at R.T. and 50 psig.

Conclusions

The use of heteropoly acids as anchoring agents for homogeneous catalysts provides a general method for preparing 'heterogenized' homogeneous catalysts using preformed complexes. The technique is very general in that there appears to be no limitation on the types of ligand present in the complex. These anchored catalysts are stable in a range of solvents with little, if any, loss of the complex during the reaction. They are generally as active and selective as the homogeneous analogs and are particularly effective with the chiral complexes in use today. These catalysts have been used for multiple batch and continuous hydrogenations with TON's of 10,000 or more and TOF's as high as 660 hr^{-1}. While most of the work done to date has been involved with anchored rhodium complexes, some work has also been done with ruthenium species. Work is presently underway on extending this technology for use with other metal complexes.

References

1. Collman, J.P.; Hegedus, L.S.; Cooke, M.P.; Norton, J.R.; Dolcetti, G.; Marquardt, D.N. *J. Amer. Chem. Soc.* **1972**, *94*,1789.
2. Dumont, W.; Poulin, J.-C.; Daud, T.P.; Kagan, H.B. *J. Amer. Chem. Soc.* **1973**, *95*, 8295.
3. Vankelecom, I.F.J.; Jacobs, P.A. In *Chiral Catalyst Immobilization and Recycling;* DeVos, D.E.; Vankelecom, I.F.J.; Jacobs, P.A., Eds.; Wiley-VCH: New York, NY, 2000, pp 19-80.
4. Bayston, D.J.; Polywka, M.E.C. In *Chiral Catalyst Immobilization and Recycling;* DeVos, D.E.; Vankelecom, I.F.J.; Jacobs, P.A., Eds.; Wiley-VCH: New York, NY, 2000, pp 211-234.
5. Hartley, F.R. *Supported Metal Complexes*; Reidel: Dordrecht, 1985.
6. Keim, W.; Dreissen-Holscher, B. In *Handbook of Heterogeneous Catalysis*; Ertl, G.; Knozinger, H.; Weitkamp, J., Eds.; Wiley-VCH: New York, NY, 1977, pp231-240.
7. Lindmer, E.; Schneller, T.; Auer, F.; Mayer, H.A. *Angew. Chem., Int. Ed.* **1999**, *38*, 2155.
8. See, for instance: Raynor, S.A.; Thomas, J.M.; Raja, R.; Johnson, B.F.G.; Bell, R.G.; Mantle, M.D. *Chem Commun. (Cambridge)* **2000**, (19), 1925; DeRege, F.M.; Morita, K.; Ott, K.C.; Tumas, W.; Broene, R.D. *Chem. Ind. (Dekker)* **2001**, *82 (Catalysis of Organic Reactions)*, 439 and references cited therein.
9. Augustine, R.; Tanielyan, S. *Chem. Commun. (Cambridge)*, **1999**, 1257.
10. Tanielyan, S.K.; Augustine, R.L. US Patents 6,005,148 (1999) and 6,025,295 (2000).
11. Tanielyan, S.K.; Augustine, R.L. PCT Int. Appl., WO-9828074: *Chem. Abstr.* **1998**, *129*, 109217.
12. Tanielyan, S.K.; Augustine, R.L. *Chem. Ind. (Dekker)* **1998**, *75 (Catalysis of Organic Reactions)*, 101.
13. Augustine, R.L.; Tanielyan, S.K.; Anderson, S.; Yang, H.; Gao, Y. *Chem. Ind. (Dekker)* **2000**, *82 (Catalysis of Organic Reactions)*, 497.
14. Goel, P.; Anderson, S.; Nair, J.; Reyes, C.; Gao, Y.; Tanielyan, S.; Augustine, R. *Chem. Ind. (Dekker)* **2002**, *(Catalysis of Organic Reactions)*, In Press.

15. Augustine, R.L.; Tanielyan, S.K. *Chem. Ind. (Dekker)* **2002**, *(Catalysis of Organic Reactions)*, In Press.
16. Benincori, T.; Cesarotti, E.; Piccolo, O.; Sannicolo, F. *J. Org. Chem.* **2000**, *65*, 2043.
17. Burk, M.J.; Gerlach, A.; Semmeril, D. *J. Org. Chem.* **2000**, *65*, 8933.

Chapter 3

Strategies for Improving Enantiomeric Purities via the Tandem Use of Mirror-Image Catalysts and Controlling Reversible Epimerizations

J. W. Faller, A. R. Lavoie, and N. Sarantopoulos

Department of Chemistry, Yale University, New Haven, CT 06520

Racemic catalysts can be made to yield enantiomerically enriched products via deactivation of one enantiomer. An alternative to this "chiral poisoning" is the use of an asymmetric transformation wherein a single enantiomer or diastereomer can be obtained upon crystallization. In some cases both enantiomers of a catalyst can be used in tandem to increase the ee of a product obtained in modest ee.

Many approaches to catalyst design use steric interactions to relay the chirality of an asymmetric catalyst to the substrate. We have been particularly interested in the application of differences in electronic asymmetry to influence enantioselectivity of catalysts (1-5). Particularly effective ligands have been those involving a phosphine and phosphine oxide donor set. In this context we have found that the ligand BINPO or BINAP(O), the monoxide of BINAP has provided a particular effective ligand for the preparation of a catalyst for the asymmetric Lewis acid catalyzed Diels-Alder reaction. Recently, we reported that enantiomerically pure [(R_{Ru},S)-*p*-cymeneRuCl(BINPO)]SbF_6 was a highly enantioselective Lewis-acid precatalyst for the asymmetric Diels-Alder reaction between methacrolein and cyclopentadiene (see Figure 1)(*1*). In fact, when this reaction was carried out at -78 • C, (*S*)-(+)-exo-2-methylbicyclo[2.2.1]hept-5-ene-2-carboxaldehyde was obtained in high conversion with de = 93 % and with an ee = 99 %.

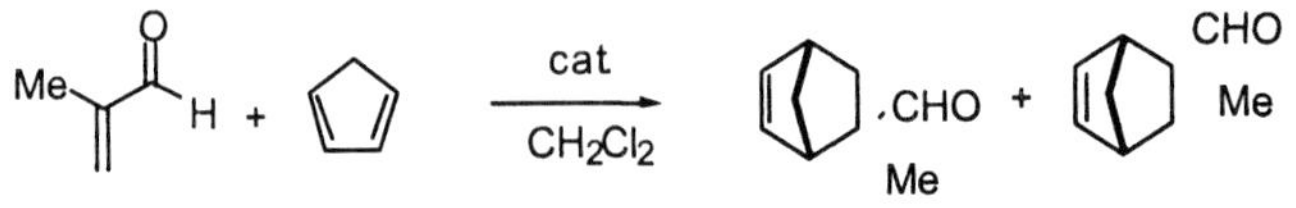

Figure 1. A Diels-Alder reaction.

The metal center in the chloro precursor is chiral by virtue of the cymene ligand, the chloro, the phosphorus and the oxygen donors. Two diastereomers are possible with the chirality at the metal and that of the ligand, but the steric interactions in the (*S*)-BINPO ligand make one diastereomer predominant (Figure 2).

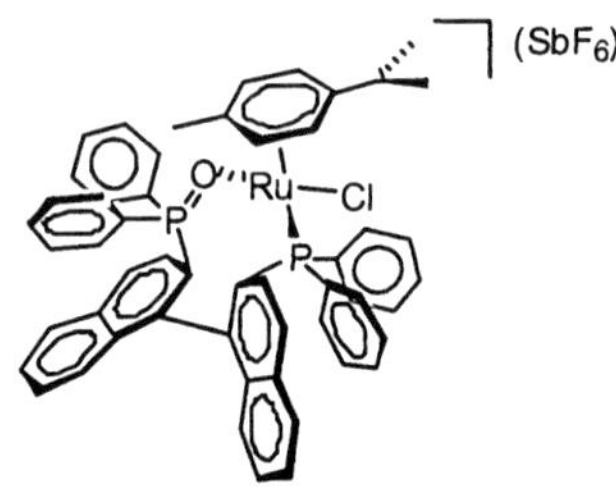

Figure 2. [(R_{Ru},S)-p-cymeneRuCl(BINPO)]SbF_6

Removal of the chloride from the precursor with $AgSbF_6$ yields a double charged Lewis acid that can bind and activate the methacrolein towards reaction with the diene. The electronic asymmetry at the metal affects the electron density in the *d*-orbitals of the metal and serves to orient the aldehyde so that attack can only occur on a specific enantioface of the olefin. Although the enantioselectivity cannot be attributed solely to the electronic asymmetry, one should note that the analogous Lewis prepared from BINAP yields only an ee of 19%, compared to 99%, under similar reaction conditions.

Chiral poisoning

The osmium analogue, [(R_{Ru},*S*)-*p*-cymeneOsCl(BINPO)]SbF_6, offers similar reactivity and enantioselectivity. Both of these are precious metal catalysts with osmium being much more expensive. The term precious metal may actually be misleading because the ligands are often more expensive than the metal. The chiral multiplication possible wherein a single catalyst molecule can transmit chirality to a large number of product molecules can make the use of these precious metals and ligands economical for practical applications. Sometimes enantiopure ligands can be synthesized from the chiral pool, but often there is a resolution step that adds to the cost of the ligand. Thus, the racemic ligand is usually substantially less expensive than the enantiomerically pure ligand. An attractive concept is the use of the racemic, but still relatively expensive, ligand in the catalyst and adding an inexpensive modifier to yield a catalyst that will produce an enantiomerically enriched product. This approach has been used successfully in a number of cases (5-16), particularly with hydrogenations, but we have recently found that it can be used with Lewis acid catalysis of the Diels-Alder reaction (17).

As shown in Figure 3, an ideal situation is that the additive selectively binds to only one of the enantiomers of the catalyst leaving the other free to catalyze the reaction.

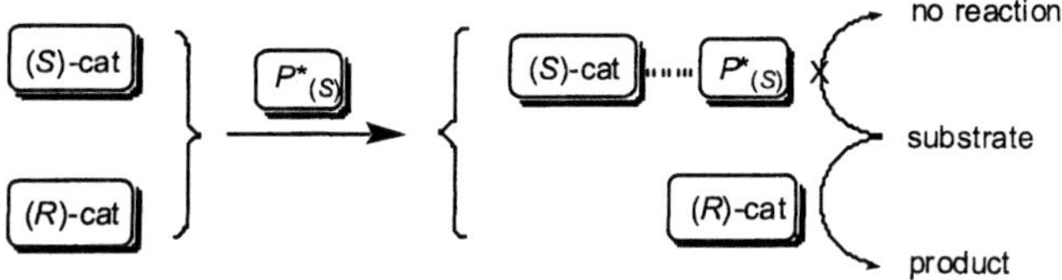

Figure 3. A selective poison or inhibitor for the (S) catalyst, $P^*_{(S)}$, allows the (R)-catalyst to produce a chiral nonracemic product.

In the case of addition of L-proline to the catalyst prepared from [(p-cymeneRuCl(BINPO)]SbF_6 prepared from racemic BINPO, one might anticipate that one enantiomer of the catalyst would be preferentially poisoned and the other would produce an enantiomeric excess in the product, as shown in figure 4. In fact use of this poison produced (*S*)-(+)-exo-2-methylbicyclo[2.2.1]hept-5-ene-2-carboxaldehyde with an ee ~ 60 %. Since the (*S*)-product has been shown to be formed preferentially by the [(R_{Ru},*S*)-*p*-cymeneRu(BINPO)(H_2O)](SbF_6)$_2$, it follows that the [(S_{Ru},*R*)-*p*-cymeneRu(BINPO)(H_2O](SbF_6)$_2$ was inhibited.

This is an idealized view, since the poison will actually bind to both enantiomers of the ruthenium complex, although preferentially to one of them. Hence one enantiomer is deactivated more that the other. Regardless, moderately high ee's are obtainable using the racemic catalyst. L-prolinamide also selectively deactivates the (*R*)-BINPO catalyst, but apparently does so by a combination of binding, as well as displacement, of the BINPO ligand. [17]

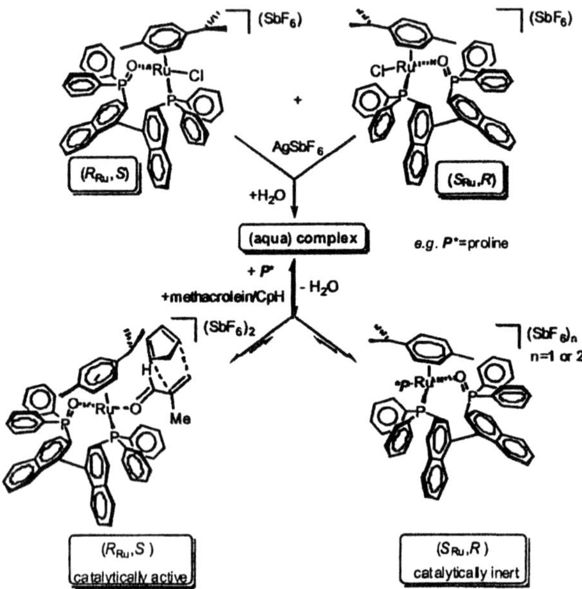

Figure 4. Chiral poisoning of a racemic mixture of a ruthenium catalyst (from reference 17).

It is not always straightforward to determine the mechanism by which a nonracemic product is obtained when an enantiopure additive is added to a racemic catalyst (18). In some cases a new catalyst can be produced that is more efficient than either enantiomer of the original catalyst. This has been termed asymmetric activation (13-16). Regardless of the nature of the interaction it is now apparent that chiral additives to racemic catalysts can provide systems that can yield products of moderate to high enantiopurity.

Dehydrogenation—a Route to Antipodes of Enantiomers Produced by Hydrogenation

The usual goal of asymmetric catalysis is high enantioselectivity. After optimization it is often desirable to enhance the scope of the system to allow the production of both enantiomers of the product in high enantiomeric purity. Employment of the other antipode of the catalyst, which is often obtained using the enantiomeric ligand in a transition metal catalyst, would be expected to yield the other enantiomer of the product. Unfortunately, in cases where only one antipode of the ligand is available, this approach is not an option.

An ideal situation in this case would be the development of a methodology that could use the same catalyst to produce either enantiomer of the product in high ee. A single catalyst can provide a route to both antipodes of a product if one considers hydrogenation and dehydrogenation using the same enantiomer of a catalyst. Although both enantiomers of amino-2-indanol are available, our studies demonstrate that both enantiomers of the product can be produced using catalysts derived from a single enantiomer of this ligand.

If a product is available in high enantiomeric purity via a transfer hydrogenation, then kinetic resolution via dehydrogenation can provide an alternative route to the other antipode. Ruthenium catalysts have been reported that effect both reactions (15, 19). A different system was developed by Wills for hydrogenations used ruthenium catalysts derived from amino alcohols (20). We have found that these systems can also be used for kinetic resolutions (21). This system allows the demonstration of the generally unappreciated notion that a single catalyst can provide a route to both antipodes of a product.

Using a (cymene)ruthenium catalyst prepared from (1*S*,2*R*)-*cis*-1amino-2-indanol, transfer hydrogenation from 2-propanol yields the (*R*)-alcohol in high enantiomeric purity (Figure 5).

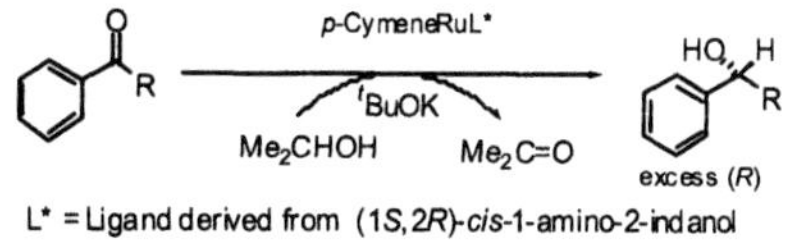

Figure 5. Asymmetric transfer hydrogenation

This reverse reaction, dehydrogenation, exploits the ability to manipulate equilibrium conditions for the reversible reaction by varying the concentration of 2-propanol or acetone (see Figure 6). In the presence of excess 2-propanol, hydrogen is transferred to a ketone substrate from 2-propanol; whereas in the presence of excess acetone, hydrogen is transferred from a chiral alcohol to the acetone.

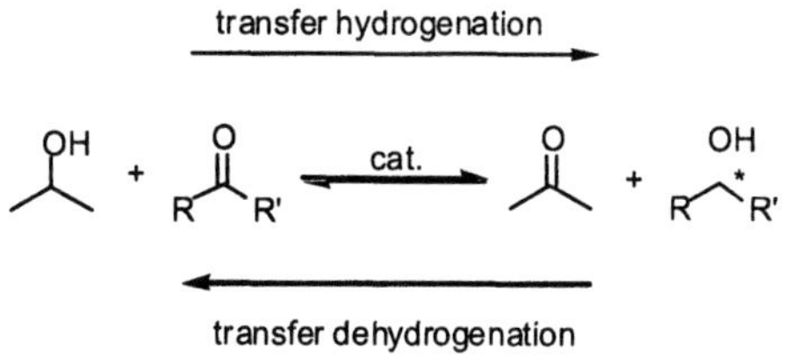

Figure 6. Transfer hydrogenation and dehydrogenation.

Since the transfer hydrogenation with a chiral catalyst and prochiral ketone produces diastereomeric transition states depending upon which prochiral faces of the ketone are involved, the rate of production of the (*R*) and (*S*) are different. In this case an enantiomeric excess of the (*R*)-enantiomer is produced (see Figure 7).

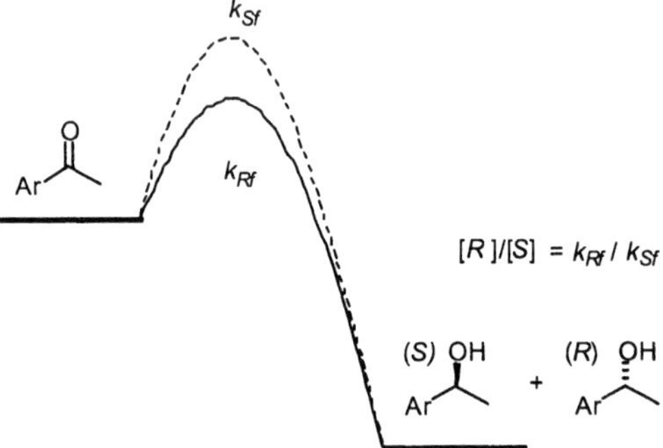

Figure 7. Transfer hydrogenation via diastereomeric transition states.

In the reverse reaction, that presumably involves a similar transition state, the thermodynamics can favor dehydrogenation of the chiral alcohol when acetone is in excess and 2-propanol is not available. Thus one can reduce the ketone to the alcohol using a conventional reducing agent to product the racemic alcohol and than essentially run the reaction backwards.

Given that the relative difference in rates (taken as a ratio=$k_{fast}/k_{slow}=k_{rel}$) is significant, then one enantiomer of the alcohol can selectively react with the chiral catalyst and induce kinetic resolution. In an ideal case, an ee = 100% could be obtained with a conversion of 50%. In reality k_{rel} >5 can produce reasonable enantiomeric purities (22, 23). Thus, an asymmetric transfer hydrogenation catalyst should be capable of dehydrogenation, as well as hydrogenation, and should be able to be used in this manner (see Figure 8). In our example, the transition state that involves the (*R*)-alcohol is lower in energy; hence, it is dehydrogenated to ketone faster than the (*S*)-alcohol. This leaves the (S)-alcohol in excess upon dehydrogenation.

Phenethyl alcohol, tetralol, and indanol were kinetically resolved by this method in 90, 97, and 99% ee at 54, 51, and 65% conversion respectively. (21)

Tandem Hydrogenation/Dehydrogenation with Mirror Image Catalysts

The optimization of a new chiral catalyst often proceeds via modification of chiral ligands in order to improve the enantioselectivity. After *tuning* the

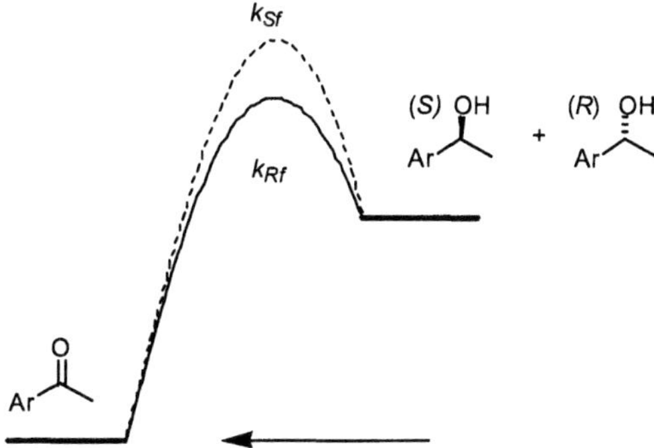

Figure 8. Transfer dehydrogenation via diastereomeric transition

system by variations in electronic effects, steric effects, or reaction conditions one hopefully is able to achieve a catalyst that provides a product of acceptable enantiomeric purity. Nevertheless, even after extensive tuning there is often a point of diminishing returns wherein only modest improvement in product ee is obtained. Relevant to investigation of asymmetric transfer hydrogenation catalysis, we have developed a general strategy which will allow for one to obtain ee's in excess of the original asymmetric hydrogenation.

This approach exploits the reversibility of the aryl ketone/alcohol equilibrium. As shown above, if the (*R*)-alcohol were produced in excess, dehydrogenation with the same catalyst would selectively dehydrogenate the (*R*)-alcohol and the ee of the remaining alcohol would decrease. On the other hand, if the reverse reaction used the mirror image catalyst, i.e., the one prepared from (1*R*,2*S*)-*cis*-1amino-2-indanol, instead of (1*S*,2*R*)-*cis*-1amino-2-indanol, the (*S*)-alcohol would be selectively dehydrogenated. This would increase the ee of the remaining product.

Figure 9 shows how the ee would be expected increase using the tandem approach if an initial ee reflects the k_{rel} for the dehydrogenation with the enantiomeric catalyst. Although high ee's could be obtained by starting with the racemic alcohol and running the kinetic resolution to high conversion, the overall yield would be low. Using the tandem approach, one gets a head start on the kinetic resolution and the product yield is higher.

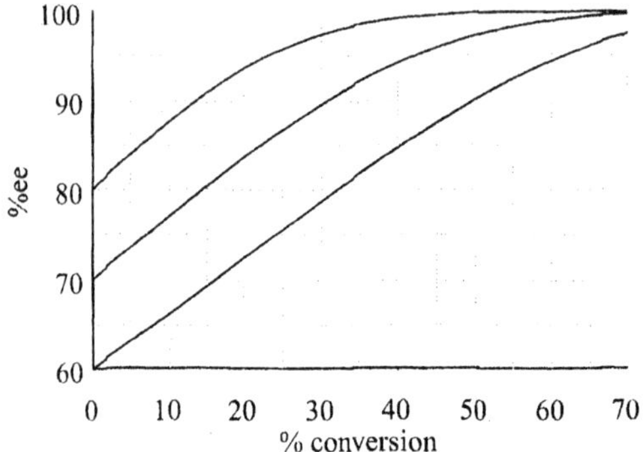

Figure 9. Enhancement of ee by kinetic resolution of a nonracemic substrate (adapted from ref 24).

For example, the ee of the 2'-acetonaphtone can be increased from an originally obtained 89% ee to 97% with a 25% conversion in kinetic resolution (See Figure 10).

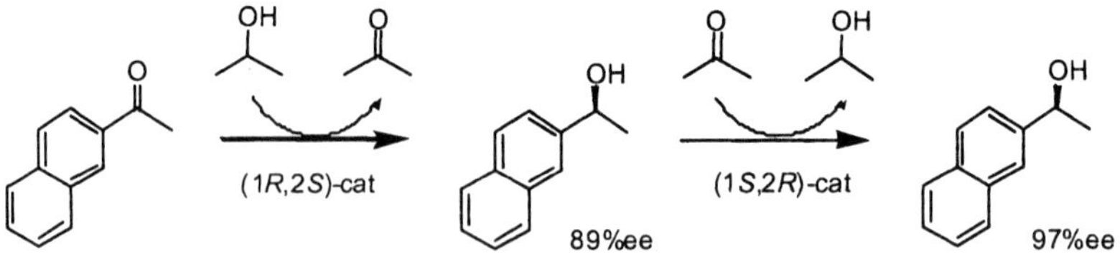

Figure 10. Tandem enhancement of ee.

Crystallization-Induced Asymmetric Transformation

When chiral nonracemic organometallics are desired one generally utilizes chiral ligands and may consider a conventional resolution procedure. In certain cases crystallization can control the stereochemistry at the metal. The following studies illustrate some of these unusual effects. These transformations have the potential of being used in the stereoselective formation of carbon-carbon bonds in organic synthesis. Stoichiometric(25) allylic alkylation is an important reaction that provides stereocontrolled formation of a carbon-carbon bond, although catalytic methods(26) are preferred. Stereoselective alkylation of allylic substrates has been obtained with molybdenum (27) and tungsten (28) complexes. Clearly, the structure of the metal complex used for the allylic alkylation reaction dictates the regio- and enantioselectivity that is experimentally observed. The presence of dynamic equilibria between different isomers of the metal-containing intermediates and these rates relative to the rate of alkylation can affect the selectivity.

Crystallization of (η^3-crotyl)Mo$(CO)_2$(diphos)Cl from methylene chloride yields dark red crystals. An X-ray crystallographic analysis showed that the solid state stereochemistry at the metal in this complex (Figure 11) is consistent with the structure of the previously synthesized (η^3-allyl)Mo$(CO)_2$(diphos)Cl (29). If one considers the η^3-crotyl to occupy a single site of an octahedral complex considered as an apical position, the two carbonyls occupy *cis* equatorial positions and diphos has one phosphorus atom trans to the axially coordinated crotyl. The other phosphorus atom of the diphos is trans to one of the carbonyls and the remaining equatorial site is occupied by the chloride. This arrangement of ligands produces a chiral metal center. Compared to the allyl analog, another element of chirality is introduced by adding a methyl substituent to the allyl; hence, enantiomers and diastereomers are possible. This is not encountered in (η^3-allyl)Mo$(CO)_2$(diphos)Cl, for which only the two enantiomers are observed. In this allyl case the compound crystallizes as a racemate with both enantiomers from the metal chirality present in a given crystal. For the crotyl derivative spontaneous resolution of the two enantiomers of (η^3-crotyl)Mo$(CO)_2$(diphos)Cl occurs upon crystallization and a conglomerate is formed. Thus, each crystal contains only one enantiomer.

A large crystal was cut to yield a portion of suitable size for X-ray crystallography and the remaining portion was dissolved to obtain the optical rotation and CD spectra. This showed that the $(+)_{589}$-(η^3-crotyl)Mo$(CO)_2$(diphos)Cl had the (*S*) configuration at the crotyl carbon and an (*A*) configuration at the metal as shown in Figure 11.

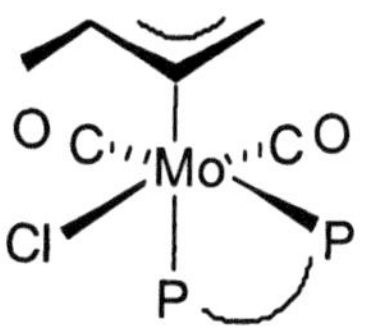

Figure 11. $(+)_{589}$-(S)-(A)-η^3-crotyl)Mo$(CO)_2$(diphos)Cl

As found for allyl complex, in solution there is a rapid interconversion between the *A* and *C* chiral center at the metal. This can be viewed as rotation of P,P,Cl ligand set around an axis toward to the metal. This is shown for the allyl complex in Figure 12.

Figure 12. Metal centered chirality interconversion.

This interconversion would produce diastereomers in each of the crotyl complexes as shown in Figure 13. This interconversion is rapid and occurs with a barrier < 11 kcal/mol.

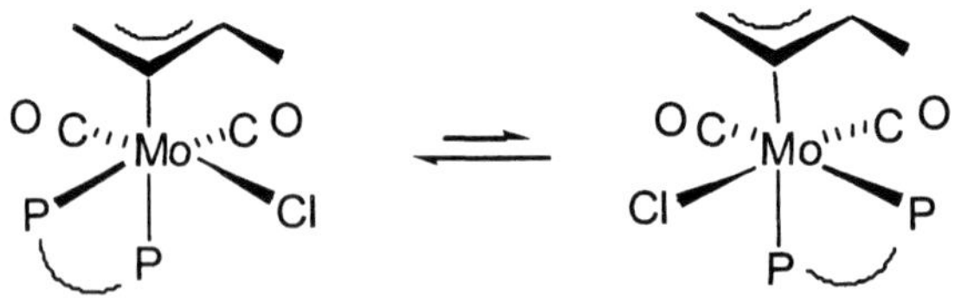

Figure 13. Diastereomer interconversion (epimerization)

A significantly slower process involves interconversion of the chirality centered on the crotyl ligand. This occurs via an η^3-η^1-η^3 process that allows a different enantioface of the crotyl to bind to the metal. This is illustrated in Figure 14.

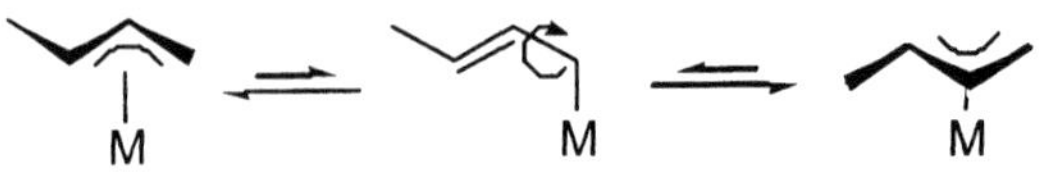

Figure 14. An η^3-η^1-η^3 –crotyl process.

In the spontaneous resolution the rates of interconversion of diastereomers are fast relative to that of crystallization and so the preferred diastereomer in the solid is found and pure enantiomers are found in each crystal. This is illustrated in Figure 15.

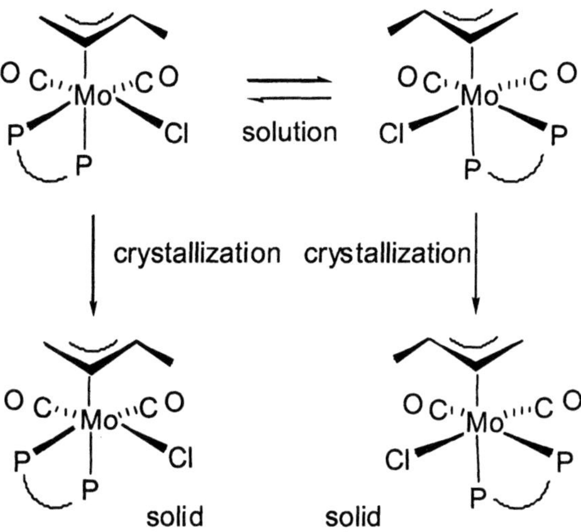

Figure 15. Spontaneous resolution

The individual crystals can be separated and harvested and those with same rotation combined to yield enantiomerically pure complexes. The crotyl interconversion is slow and occurs with a half-life of 3.8 h at 20 °C. This would allow sufficient time to carry out a reaction on the metal complex before racemization occurred.

The mechanical separation of the crystals can be avoided by using a chiral bisphosphine. The (*S*,*S*)-chiraphos derivative epimerizes at the crotyl with a half life of 3.5 h. Although diastereomers exist in solution there is little selectivity and the equilibrium constant is ~1. Upon crystallization, however, only a single enantiomer (diastereomer) is found in the crystals (see Figure 16).

As the crystallization proceeds the equilibrium in solution keeps replacing the isomer that crystallizes from solution and eventually the entire complex is converted into the same isomer.

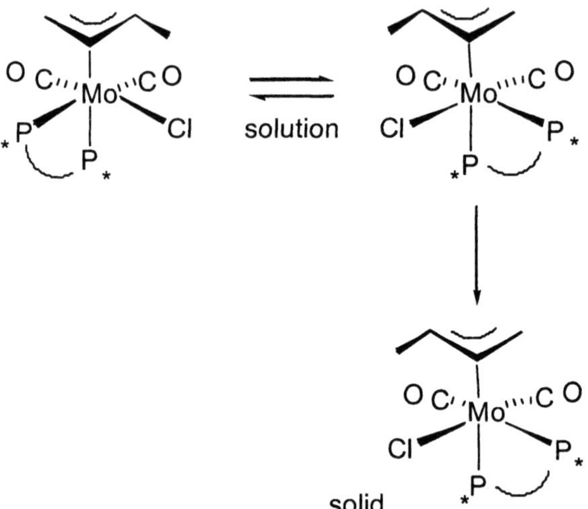

Figure 16. Crystallization-Induced Asymmetric Transformation

These methods offer alternative strategies to conventional approaches for producing asymmetric organometallics.

Acknowledg ment. We thank the National Science Foundation (Grant CHE0092222) for support of this research.

References

1. Faller, J. W.; Grimmond, B. J.; D'Alliessi., D. G., *J. Am. Chem. Soc.*, **2001**, *123*, 2525.
2. Faller, J. W.; Lavoie, A. R. *J. Organometal. Chem.* **2001**, *630*, 17.
3. Faller, J. W.; Grimmond, B. J. *Organometallics* **2001**, *20*, 2454.
4. Faller, J. W.; Parr, J. *Organometallics* **2001**, *20*, 697.
5. Alcock, N. W.; Brown, J. M.; Maddox, P. J. *J. Chem. Soc., Chem. Commun.*, **1986**, 1532.
6. Brown, J. M.; Maddox, P. *Chirality*, **1991**, 3, 345.
7. Maruoka, K.; Yamamoto, H. *J. Am. Chem. Soc.*, **1989**, 111, 789.
8. Faller, J. W.; Parr, J. *J. Am. Chem. Soc.* **1993**, 115, 804.
9. Faller, J. W. Tokunaga, M. *Tetrahedron Lett.*, **1993**, 34, 7359.
10. Faller, J. W.; Mazzieri, M. M.; Nguyen, J. T.; Parr, J.; Tokunaga, M. *Pure & Appl. Chem.*, **1994,** 66, 1463.
11. Faller, J. W.; Sams, D. W. I.; Liu, X. *J. Am. Chem. Soc.*, **1996,** 118, 1217.

12. Faller, J.W.; Liu, X. *Tetrahedron Lett.*, **1996**, 37, 3449.
13. Sablong, R.; Osborn, J. A.; Faller, J. W. *J. Organometal. Chem.*, **1997**, 527, 65.
14. Matsukawa, S.; Mikami, K. *Tetrahedron-Asymmetry* **1997**, *8*, 815.
15. Doucet, H.; Ohkuma, T.; Murata, K.; Yokozawa, T.; Kozawa, M.; Katayama, E.; England, A. F.; Ikariya, T.; Noyori, R. *Angew. Chem. Int. Ed.* **1998**, *37*, 1703.
16. Mikami, K.; Korenaga, T.; Ohkuma, T.; Noyori, R. *Angew. Chem. Int. Ed.* **2000**, 39, 3707.
17. Faller, J. W.; Lavoie, A. R. Grimmond, B. J. *Organometallics* **2002**, *21,* 1662.
18. Blackmond, D. G. *Accts. Chem. Res.* **2000**, *33*, 402.
19. Hashiguchi, S.; Fujii, A.; Haack, K. -J.; Matsumura, K.; Ikariya, T.; Noyori, R. *Angew. Chem. Int. Ed. Engl.*, **1997**, *36*, 288.
20. Palmer, M.; Walsgrove, T.; Wills, M. *J. Org. Chem.* **1997**, *62*, 5226.
21. Faller, J. W.; Lavoie, A. R. *Organic Letters* **2001**, *3*, 3703.
22. Kagan, H. B.; Fiaud, J. C. *Top. Stereochem.* **1988**, *18*, 249.
23. Keith, J. M.; Larrow, J. F.; Jacobsen, E.N. *Adv. Synth. Catal.* **2001**, *5*, 343.
24. Faller, J. W.; Lavoie, A. R. *Organometallics* **2002**, *21*, 2010.
25. Faller J. W.; Chao K.-H. *Organometallics* **1984**, *3*, 927.
26. Trost B. M.; Van Vranken D. L. *Chem. Rev.* **1996**, *96*, 395.
27. Trost B. M.; Hachiya I. *J. Am. Chem. Soc.* **1998**, *120*, 1104.
28. Lloyd-Jones G. C.; Pfaltz A. *Angew. Chem. Int. Ed. Engl.* **1995**, *34*, 462.
29. Faller J. W.; Haitko D. A.; Adams R. D.; Chodosh D. F. *J. Am. Chem. Soc.* **1979**, *101*, 865.

Chapter 4

Asymmetric Organocopper Chemistry: Cu-Catalyzed Conjugate Addition and Allylic Substitution

Alexandre Alexakis

Department of Organic Chemistry, University of Geneva, 30 quai Ernest Ansermet, Genève 4, Switzerland CH–1211

This review deals with the most recent developments in the asymmetric conjugate addition and allylic substitution. For the conjugate addition, the best enantioselectivities (>99%) have been attained with dialkylzinc reagents and 0.5-2% CuX and 1-4% of a chiral trivalent phosphorus ligand. The γ-allylic substitution can be achieved equally well with, either dialkylzinc or Grignard reagents, and the same catalysts.

Oragnocopper chemistry is a standard synthetic tool nowadays (1). There are thousands of natural products, which have been synthesized using this chemistry, at least in one step. The most popular reactions are 1) the conjugate addition, 2) the substitution, and particularly the allylic substitution, 3) the cleavage of epoxides, and 4) the carbocupration.

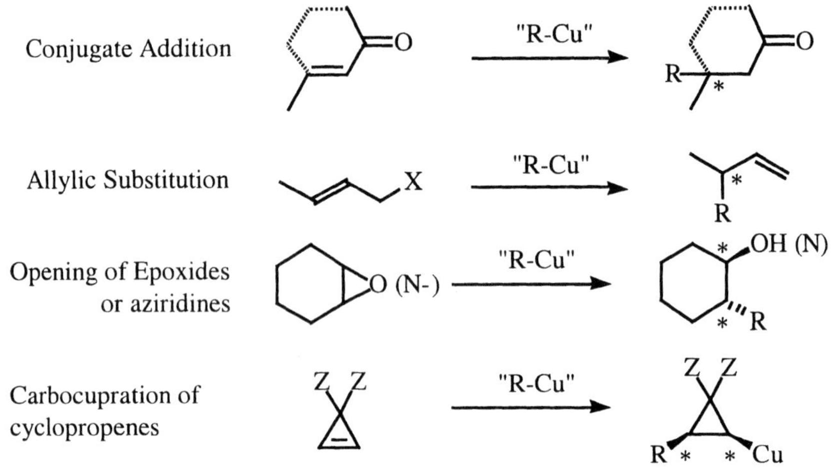

Conjugate Addition "R-Cu"

Allylic Substitution X "R-Cu" R

Opening of Epoxides or aziridines O (N-) "R-Cu" OH (N) R

Carbocupration of cyclopropenes Z Z "R-Cu" Z Z R Cu

Scheme 1

Since all these topics generate new stereogenic centers, several solutions have been explored to control the enantioselectivity. It is, however, only recently that the copper-catalyzed reactions met a breakthrough, particularly for the first three reactions.

The conjugate addition

There are many ways to bring the asymmetric information. However, only the last approach allow for a catalytic use of chiral source.

1° Covalent chiral auxiliaries

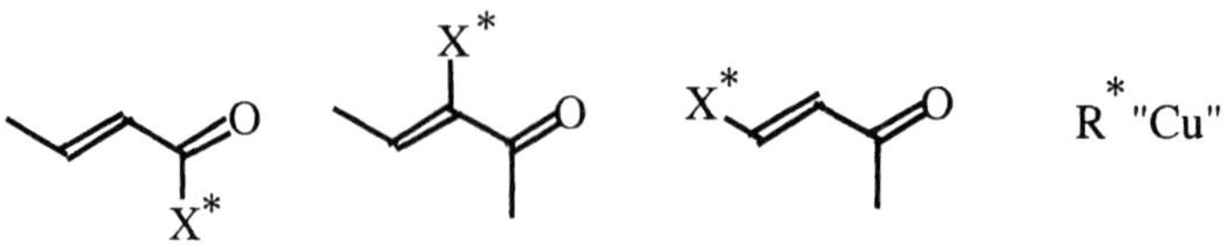

O X* X* O X* O R* "Cu"

2° Chiral ligands

$$RCu + Li\text{-}X\text{-}R^* \longrightarrow R\text{-}\overset{Li}{Cu}\text{-}X\text{-}R^* \quad \textit{"heterocuprates"}$$

$$R''Cu'' + L^* \longrightarrow R''Cu'',L^* \quad \textit{"external ligand"}$$

Scheme 2

For years, the covalent chiral auxiliary approach was the preferred one, and only few articles dealt with chiral ligands (2). Although the heterocuprate way was considered as the most practical one, it is the external ligand that brought the solution to the problem (3). A first notable success was reported in 1985 by Leyendecker who obtained 92% ee on conjugate addition to chalcone, with a stoichiometric proline-derived ligand (4). Later on, Alexakis introduced, in 1991, the use of chiral trivalent phosphorus ligands and demonstrated their efficiency (5). For catalytic amount of copper salt, and therefore chiral ligand, the same authors showed that dialkylzinc reagents were more appropriate than the classical use of Grignard reagents (6). It followed that in the last 6-7 years, a tremendous effort has been put, by more and more authors, in disclosing more and more efficient chiral ligands (>350 !), most of them bearing a phosphorus atom.

The general reaction scheme is the following:

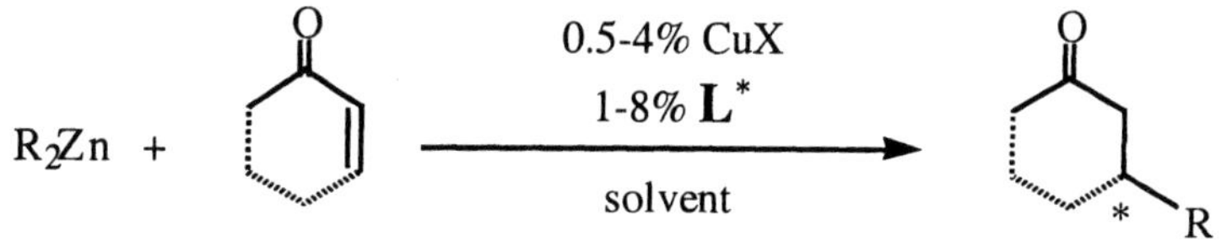

R_2Zn + [enone] — 0.5-4% CuX, 1-8% $\mathbf{L}^*$, solvent → [3-R-ketone, *, R]

Scheme 3

Dialkylzinc reagents react very sluggishly with enones, even in the presence of small amount of copper salt. A dramatic increase of reactivity is observed upon addition of the chiral ligand. This ligand *accelerated catalysis* is the cornerstone of this reaction, and is operative in non-coordinating (toluene, dichloromethane) or slightly coordinating solvents (Ether, THF, EtOAc) (7).

In addition to the solvent, all the other parameters of the reaction have been studied more or less. Copper (I) as well as copper (II) salts have been used. The true catalytic species is Cu(I), and therefore the reduction of Cu(II) is the first step in the process (see scheme 4). For practical reasons it is often more convenient to work with Cu(II) salts ($CuOTf_2$ or $CuOAc_2$ for example). Copper (II) triflate is most often the salt of choice, although copper acetate (as hydrate) and copper thiophenecarboxylate (CuTC) show superior enantioselectivity in many cases (8).

A tentative catalytic cycle is shown in scheme 4, with a copper carboxylate. After, reduction to the Cu(I) species, a first transmetallation with dialkylzinc forms a zinc carboxylate associated with an alkyl copper. This stoichiometric species does not react easily with an enone, showing that a "higher order" cuprate species must be formed with one or more additional dialkylzincs. Such a dinuclear species should coordinate to the carbonyl of the enone by the most Lewis acidic metal, zinc. At the same time, a π complexation must occur between the enone and Cu. At that point, only one ligand should be present in this transition state, as most studies have shown little or negligible non-linear effects (9). According to the usually accepted mechanism, an oxidative addition should give rise to the formation of a putative Cu(III) intermediate, which immediately collapses, by reductive elimination, to the conjugate adduct (10). This zinc enolate may be trapped by other electrophiles than water (see below).

From this catalytic cycle it appears that only one equivalent of copper salt is needed. However, the early studies revealed that higher enantioselectivities were obtained with a 2:1 ratio of ligand to copper, (5,6) and most authors systematically used these conditions. In fact, we observed that the enantioselectivity remains roughly the same with as little as a 1.2:1 ratio. In the allylic substitution, this ratio may even go to a true 1:1 ratio.

The initially explored chiral ligands were trivalent phosphorus ligands (5,6). Although other ligand types have been disclosed, the ones based on phosphorus are the most effective. Most ligands are monodentate, but some are bidentate, either P,P or P,N. An exhaustive review shows all the ligands known until April 2002 (329 at that date), (3e) and more have been disclosed since then. It is striking to see that most phosphorus ligands are of the phosphite and phosphoramidite type. Aryl phosphines are scarce, and successful only when associated with another coordination site (11). The usual chiral diphosphines, such as BINAP, are ineffective in this reaction (12). It should be pointed out that there is not yet a general ligand for all kind of Michael acceptors. Some of the most representative phosphorus ligands are shown in scheme 5.

Other classes of ligands, without phosphorus atom, have also been studied. They are not yet as efficient, although ee's as high as 93% (22) have been achieved.

In all the above-discussed reactions, dialkylzinc reagents have been used. Only in few cases, trialkylaluminum (R_3Al) were tested, but they represent an interesting alternative (24,25,26). Among R_2Zn reagents, Et_2Zn is by far the most used. Me_2Zn is about 30 times less reactive (28), and gives lower ee's, (29) although quite high in some cases (>95%) (30). Functionalized dialkylzincs afford comparable enantioselectivities as Et_2Zn (13,15,21). The compatibility of R_2Zn with many functional groups is clearly an advantage of the methodology (31). However, it should be pointed out that diaryl or divinyl zinc reagents are scarce. There is only one report on Ph_2Zn, with low ee (23). Clearly, for the transfer of an aryl or vinyl group, the Miyaura-Hayashi methodology ($ArB(OH)_2$ and Rh catalysis) is superior, and complementary to the present one (32).

The range of Michael acceptors is quite large. Traditionally, cyclohexenone

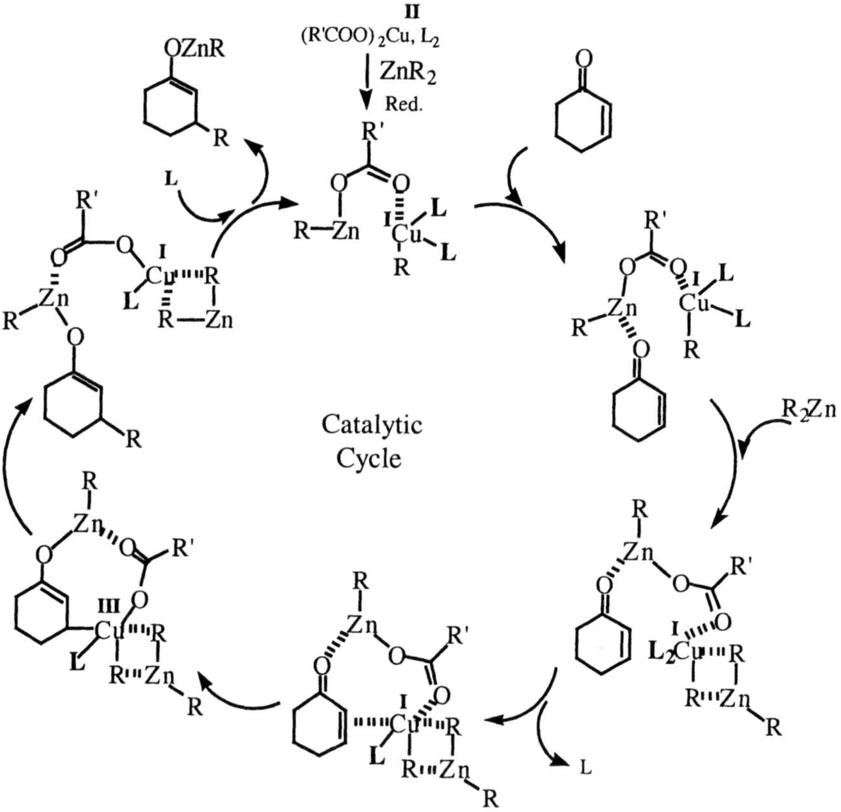

Catalytic Cycle

Scheme 4

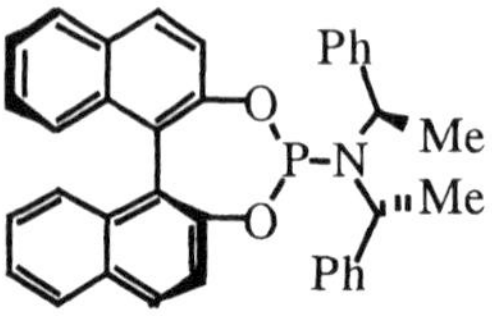

Feringa ref 13

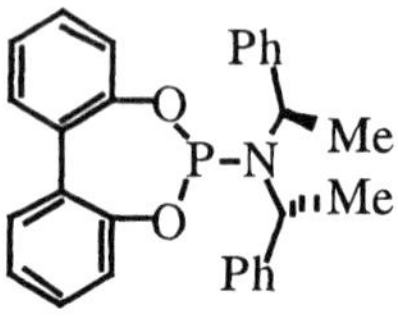

Alexakis ref 14

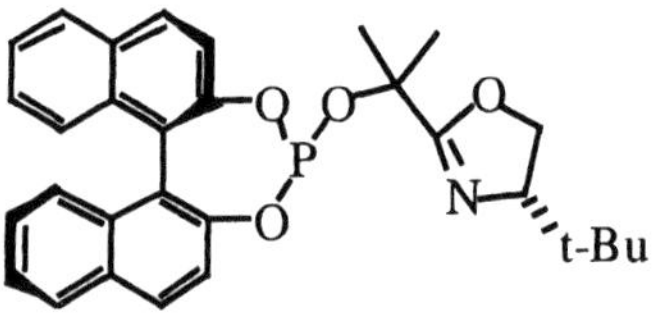

Pfaltz ref 15

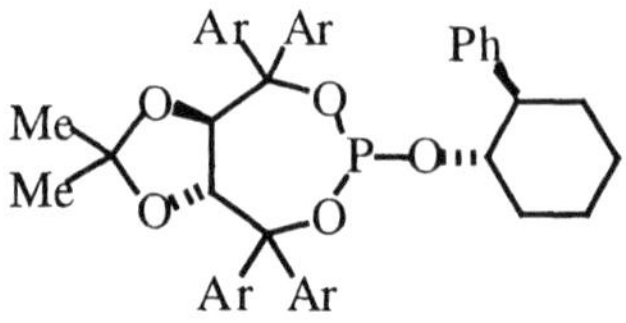

Alexakis ref 16

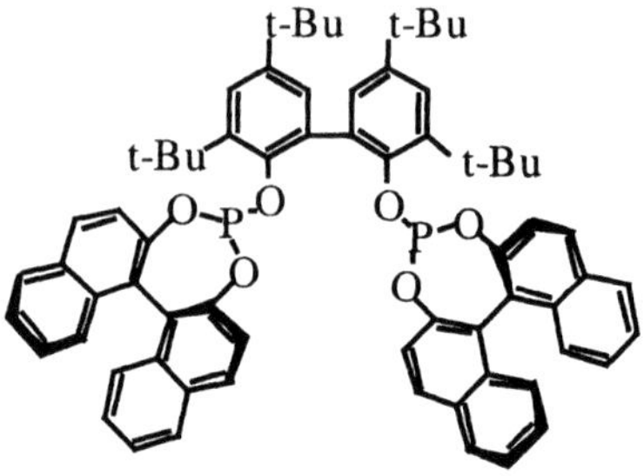

Chan ref 17

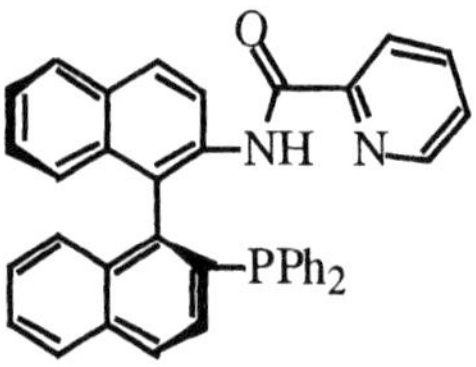

Zhang ref 18

Imamoto ref 19

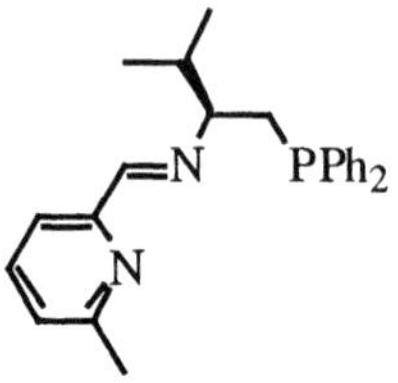

Morimoto ref 20

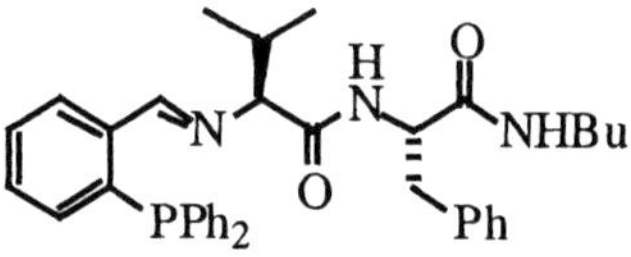

Hoveyda ref 21

Scheme 5

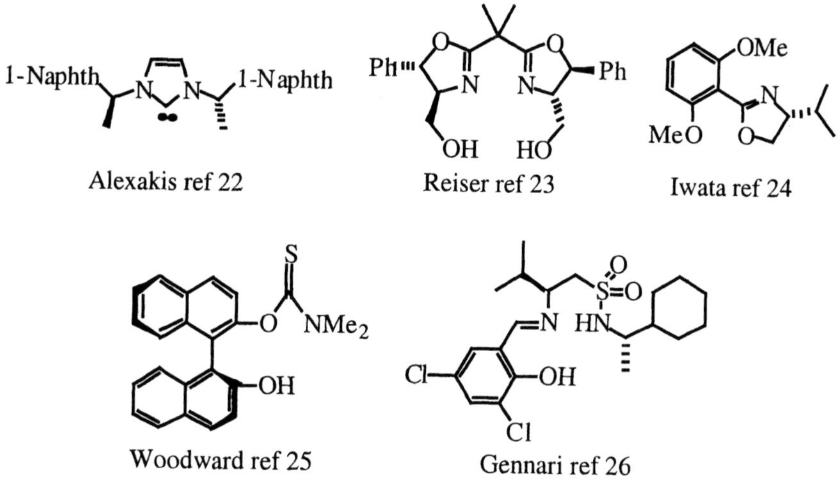

1-Naphth N N 1-Naphth

Alexakis ref 22

Ph O N O N Ph OH HO

Reiser ref 23

OMe MeO N O

Iwata ref 24

S O NMe$_2$ OH

Woodward ref 25

O S O N HN Cl OH Cl

Gennari ref 26

Scheme 6

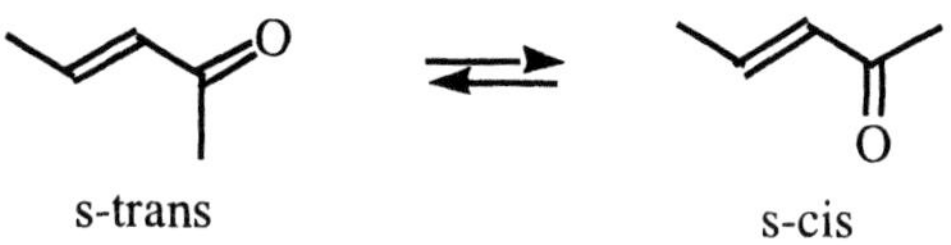

O O

s-trans

s-cis

Scheme 7

has been the substrate of choice for testing a new ligand. This cyclic enone avoids the *s-cis/s-trans* interconversion of acyclic substrates.

Most of the Michael acceptors are shown in scheme 8. Cycloheptenone and cyclooctenone behave exactly as cyclohexenone and give high ee's with the same ligands. Cyclopentenone however, is rather a flat molecule. Specific ligands have been developed especially for this substrate (21,33). Other cyclic enones include substituted cyclohexenone and cyclohexadienones (34). They give rise to efficient kinetic resolution, depending on the position of the substituent in the ring.

Many authors have equally well tested acyclic enones (25,29,35).The specific class of chalcone-type enones ($R^1 = R^2$ = Aryl) usually needs different ligands than the other ones (18, 34). Many ligands are able to bring high enantioselectivity (>95%). Of particular interest is the 15-membered macrocyclic enone, which allow the formation of (R)-Muscone, a valuable natural fragrance (25, 29).

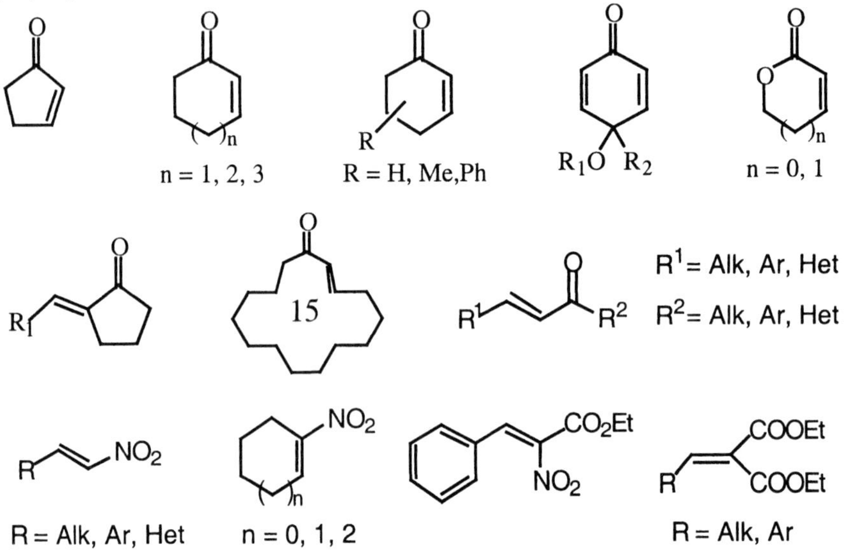

n = 1, 2, 3 R = H, Me,Ph R_1O R_2 n = 0, 1

R^1= Alk, Ar, Het
R^2= Alk, Ar, Het

R = Alk, Ar, Het n = 0, 1, 2 R = Alk, Ar

Scheme 8

Nitro-olefins are another class of excellent Michael acceptors for this reaction (12) Again, the efficient ligands are different from the previous ones (30, 36). The chiral adducts are valuable synthons, since the nitro group has successfully been transformed into amino group (by reduction) or to carbonyl groups (by a Nef reaction).

Simple α,β-ethylenic esters are not reactive enough. A good alternative is a double activation with alkylidene malonates (37). Recently, N-acyloxazolidinones have been shown to also be good ester equivalents (38).

Compared to the amount of work done recently with dialkylzincs, Grignard reagents seem to have been neglected. Historically, the first interesting levels of enantioselectivities (74%) were obtained by Lippard, in 1988, with 3% of an amidocopper catalyst (39). Later on, various copper thiolates gave moderate to good results on cyclic and acyclic enones (40-43). The best results were however obtained with external ligands, by Tomioka (44) and Sammakia (45). Since the late 90's all authors focused on the dialkylzinc procedure.

Ph NCu N Ph Lippard ref 39

N(Me)2 SCu van Koten ref 40

O N i-Pr SCu Pfaltz ref 41

O O SCu O O O Spescha ref 42

Ph Ph NMe2 Cu.LiCl O O S Ph Ph Seebach ref 43

N P(Ar)2 t-Bu O Tomioka ref 44

Ph N O Fe PPh2 Sammakia ref 45

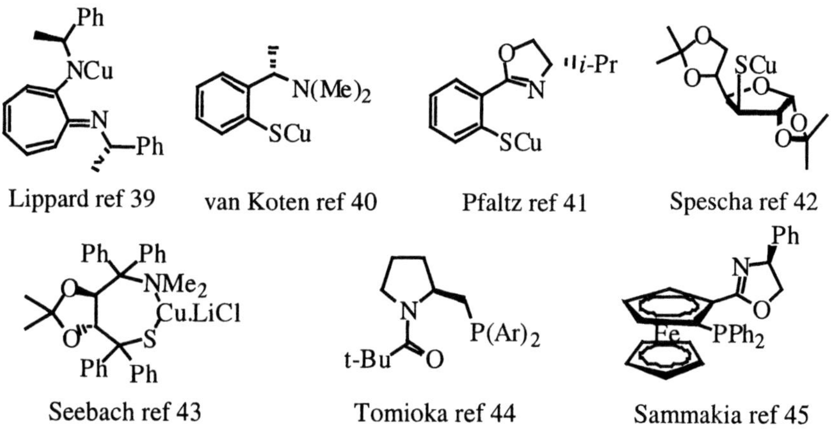

Scheme 9

The range of substrates tested with Grignard reagents is rather limited to cyclohexenone and cycloheptenone, for cyclic substrates, and chalcone and benzalacetone for acyclic ones. However, the variety of Grignard reagents is larger, with alkyl, aryl and vinyl Grignards. The enantioselectivities are in the 80-90% at best, and the catalyst loading rather high (5-15% CuX, 10-30% L*) as compared to the dialkylzinc procedure.

As seen above, all the conjugate addition reactions end up with a zinc enolate. Its reaction with an electrophile, other than simple water, could be an excellent way to quickly build more complex molecules. Despite their low reactivity, zinc enolates have been reported to react with aldehydes, (7a,28,46) with acetals (with $BF_3.Et_2O$ as additive) (47), with allylic acetates (with Pd catalysis) (7a,28,48) and with homopropargylic iodide (21), methyl iodide (49) or benzyl iodide (11d) (tenfold excess and 10 equiv. HMPA as additive).

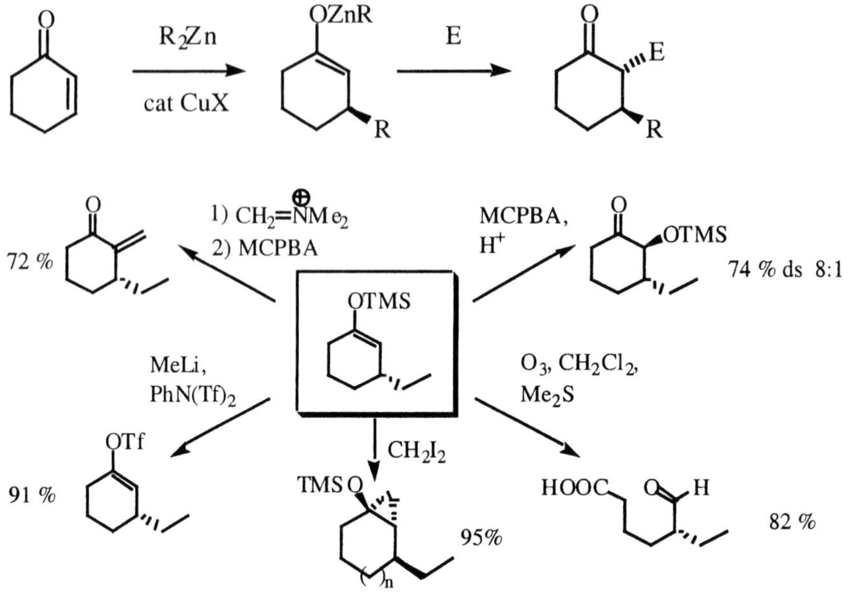

Scheme 10

Another possibility is to trap the enantiopure zinc enolate with a silylating agent (50). The resulting silyl enol ether becomes a versatile building block, able to provide many new synthons (scheme 11) (51).

The allylic substitution

This topic adds on the difficulties, as it needs the additional control of the regioselectivity. The stereochemical outcome of the allylic substitution with organocopper reagents is well established (10, 52). The reaction proceed through an S_N2 or S_N2' path with a clean *anti* selectivity. The control of the regioselectivity is usually done by the proper choice of the Cu salt, of the solvent and the additive (such as $BF_3.Et_2O$) (53). There are only two exceptions affording a clean *syn* S_N2': when the leaving group X is a secondary carbamate (54) or an o-diphenylphosphino benzoate (55).

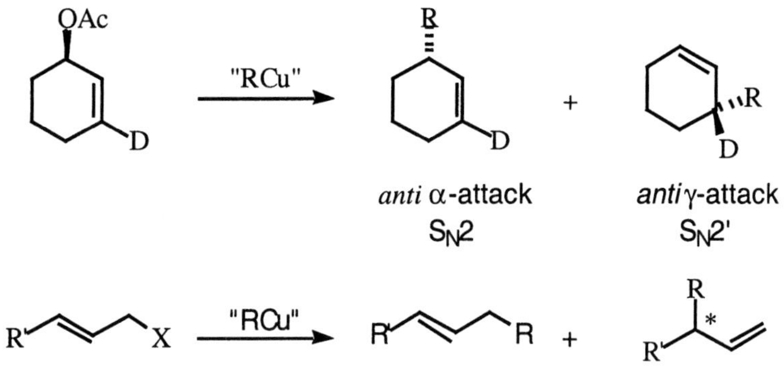

OAc

"RCu"

R

+

R

D

D

D

anti α-attack
S_N2

anti γ-attack
S_N2'

R' X "RCu" R' R + R * R'

Scheme 11

The enantioselective allylic substitution, when no stereogenic centers are on the substrate, has not been studied extensively. There are four authors who described such a reaction with a stoichiometric auxiliary, placed on the leaving group. These are chiral C2 symmetrical acetals (56), carbamates (57), oxazolinyl thioethers (58) and sulfoximines (59).

The first report on the catalytic version is due to Backvall and van Koten, in 1995 (60). They used a Grignard reagent as primary organometallics, and a copper thiolate as chiral catalyst (15%). With Alexakis, they are the only authors to report on Grignard reagents (61); all other authors have used dialkylzinc instead. Following Knochel's work, in 1999 (62) Feringa (63), Hoveyda (64), Woodward (65) and Gennari (66) have reported their results with dialkylzincs. The general equation, and the various ligands used are shown in scheme 12. The scope of the reaction is not yet very large as the substrate is concerned. Knochel's procedure is better for hindered dialkylzincs (ee's up to >95%), whereas Hoveyda's one allow for the efficient reaction with trisubstituted allylic phosphates. Alexakis has shown that ω-ethylenic Grignard reagents can be used, and treated *in situ* with Grubbs' metathesis catalyst to afford directly new chiral synthons (61b).

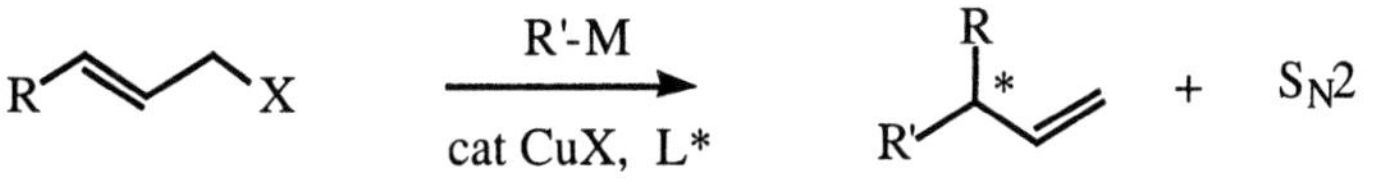

R X R'-M cat CuX, L* R * R' + S_N2

R = Alk, Ar

X = Cl, Br, OAc, $OP(O)(OEt)_2$

M = MgX, Zn

CuX = CuI, CuCN, CuOTf, $CuBr.Me_2S$,
Cu-thiophene carboxylate (CuTC)

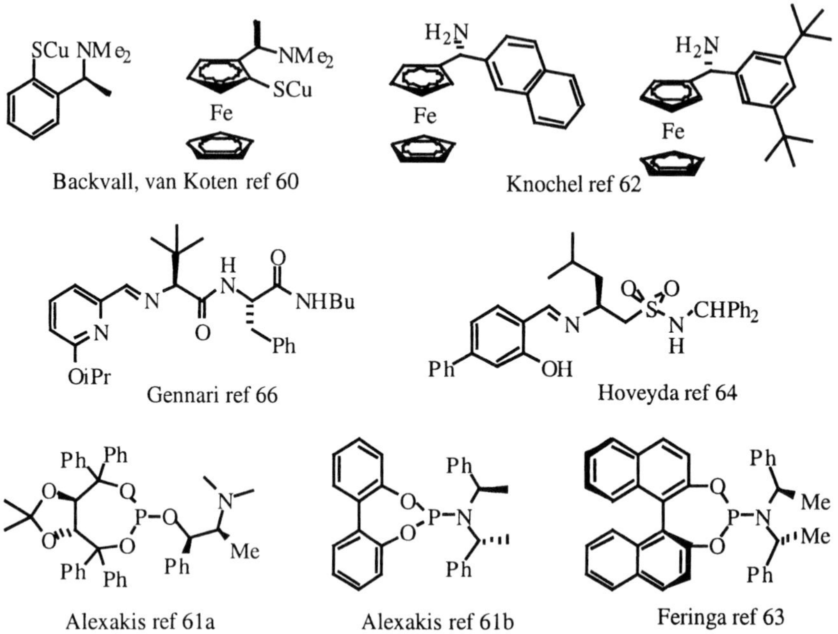

Backvall, van Koten ref 60

Knochel ref 62

Gennari ref 66

Hoveyda ref 64

Alexakis ref 61a

Alexakis ref 61b

Feringa ref 63

Scheme 12

Reactions with epoxides and aziridines

Organocopper chemistry is also useful for smooth ring opening of epoxides (67). Meso-type epoxides, such as cyclohexene oxide, are the substrates of choice for studying enantioselective versions (68). Until now, there is no report on such a copper-catalyzed reaction with ee's over 20-30%. Interestingly, the asymmetric ring opening of meso aziridines is quite efficient (15-91% ee). Among the many catalysts tested, the following copper carboxylate gave the highest enantioselectivity with yields. It should be pointed out that the amount of catalyst is crucial: with 10%, the ee is only 55%, and with 20%, it jumps to 77%, the highest ee 91% being attained with 30% catalyst (69).The same reaction on cyclohexene oxide afforded only 10% ee (69).

Pineschi and Feringa have studied the kinetic resolution on a more reactive class of epoxides: the α,β-unsaturated ones. They found that dialkylzinc reagents

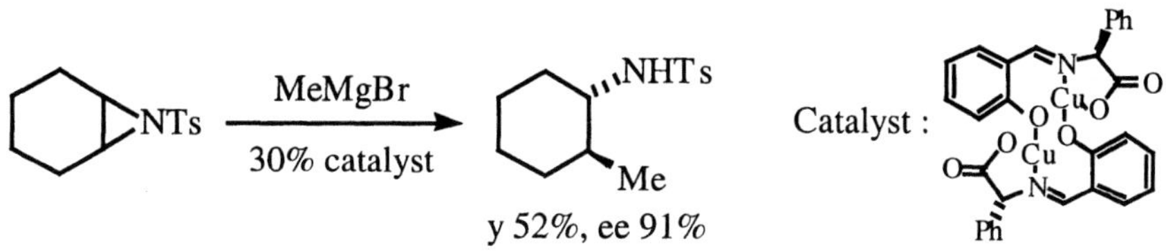

MeMgBr
30% catalyst
y 52%, ee 91%
Catalyst :

Scheme13

undergo a copper-catalyzed S_N2' type reaction. At mid-conversion, the S_N2' products may reach high ee's: 50-96% (34, 70). The reaction has been applied, with equal success, to structurally related epoxides (71).

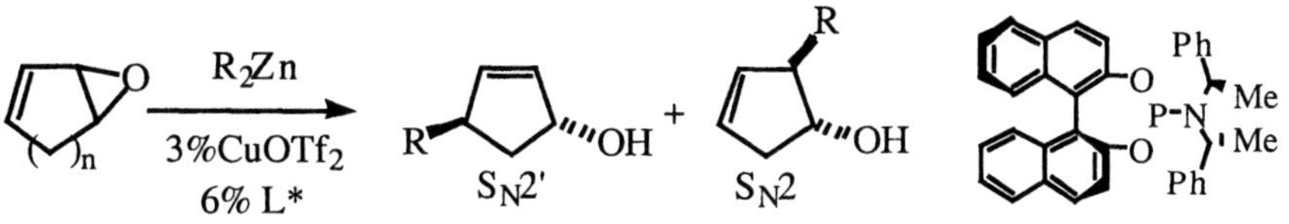

R_2Zn
3%$CuOTf_2$
6% L*
S_N2' + S_N2

Scheme 14

This short review points to the booming recent interest for the asymmetric reactions of organocopper chemistry. The synthetic potentiality of the resulting synthons makes this methodology among the most versatile for the synthesis of useful natural products, such as pharmaceuticals, fragrances etc ... Of course, further improvements are needed to enhance the enantioselectivities or the scope of these reactions, but, after 30 years of investigations, efficient solutions have finally came up.

References

1. a) Posner, G. H. *An introduction to synthesis using organocopper reagents*, Wiley Interscience, New York, **1980**. b) Taylor, R. J. K. *Oragnocopper reagents: a practical approach*, Oxford Univ. Press, **1994**. c) Ibuka, T. *Organocopper reagents in organic synthesis*, Camellia and Rose Press, Osaka, **2000**. d) Krause, N. *Modern organocopper chemistry*, Wiley-VCH, Weinheim, **2002**.
2. For a review covering the literature until 1992 see: Rossiter, B. E.; Swingle, N. M. *Chem. Rev.* **1992**, *92*, 771.

3. For other reviews covering asymmetric organocopper chemistry see: a) Alexakis, A. in *Transition Metal Catalysed Reactions*; S.-I. Murahashi, S.G. Davies, Ed.; IUPAC Blackwell Science, Oxford **1999**; p 303. b) Tomioka, K.; Nagaoka, Y. in *Comprehensive Asymmetric Catalysis*, ed. E. N. Jacobsen, A. Pfaltz, H. Yamamoto, Springer, New York **2000**, 1105. c) Sibi, M. P.; Manyem, S. *Tetrahedron* **2000**, *56*, 8033. d) N. Krause, N.; Hoffmann-Röder, A. *Synthesis* **2001**, 171. e) Alexakis, A.; Benhaim, C. *Eur. J. Org. Chem.* **2002**, 3221.
4. Leyendecker, F.; Lancher, D. *New J. Chem., 1985*, *9*, 13.
5. Alexakis, A.; Mutti, S.; Normant, J. F. *J. Am. Chem. Soc.*, **1991**, *113*, 6332-6334.
6. Alexakis, A.; Frutos, J. C.; Mangeney, P. *Tetrahedron : Asymmetry*, **1993**, *4*, 2427.
7. a) Kitamura, M.; Miki, T.; Nakano, Noyori, R. *Tetrahedron Lett.* **1996**, *37*, 5141. b) De Vries, A., H., M.; Meetsma, A.; Feringa B., L. *Ang. Chem. Int. Ed. Engl.* **1996**, *35*, 2374. c) Alexakis, A.; Vastra, J.; Mangeney, P. *Tetrahedron Lett.*, **1997**, *38*, 7745.
8. Alexakis, A.; Benhaim, C.; Rosset, S.; Humam, M. *J. Am. Chem. Soc.,* **2002**, *124*, 5262.
9. a) Zhou, Q.-L.; Pfaltz, A. *Tetrahedron* **1994**, *50*, 4467. b) Delapierre, G.; Constantieux, T.; Brunel, J.-M.; Buono, G. *Eur.J. Org. Chem.* **2000**, 2507. c) Arnold, L. A.; Imbos, R.; A. Mandoli, A.; De Vries, A. H. M.; R. Naasz, Feringa, B. L. *Tetrahedron* **2000**, *56*, 2865. d) Bennett, S. M. W.; Brown, S. M.; Cunningham, A.; Dennis, M. R.; Muxworthy, J. P.; Oakley, M. A.; Woodward, S. *Tetrahedron* **2000**, *56,* 2847.
10. For a review on mechanistic aspects of organocopper chemistry see: Nakamura, E. I.; Mori, S. *Angew. Chem., Int. Ed. Engl.* **2000**, *39*, 3750.
11. a) Hu, X.; Chen, H.; Zhang, X. *Angew. Chem., Int. Ed.* **1999**, *38*, 3518. b) Morimoto, T.; Yamaguchi, Y.; Suzuki, M.; Saitoh, A. *Tetrahedron Lett.* **2000**, *41*, 10025. c) Degrado, S. J.; Mizutani, H.; Hoveyda, A. H. *J. Am. Chem. Soc.* **2001**, *123*, 755. d) Mizutani, H.; Degrado, S. J.; Hoveyda, A. H. *J. Am. Chem. Soc.* **2002**, *124*, 779.
12. Alexakis, A.; Burton, J.; Vastra, J.; Mangeney, P. *Tetrahedron: Asymmetry* **1997**, *8*, 3987.
13. Feringa, B. L.; Pineschi, M.; Arnold, L. A.; Imbos, R.; De Vries, A. H. M. *Angew. Chem. Int. Ed.* **1997**, *36*, 2620.
14. Alexakis, A.; Rosset, S.; Allamand, J.; March, S.; Guillen, F.; Benhaim, C. *Synlett* **2001**, 1375.
15. Knöbel, A. K. H.; Escher, I. H.; Pfaltz, A. *Synlett* **1997**, 1429.
16. Alexakis, A.; Vastra, J.; Burton, J.; Benhaim, C.; Mangeney, P. *Tetrahedron Lett.* **1998**, *39*, 7869.

17. Yan, M.; Yang, L.-W.; Wong, K.-Y.; Chan, A. S. C. *Chem. Commun.* **1999**, 11
18. Hu, X.; Chen, H.; Zhang, X. *Angew. Chem. Int. Ed.* **1999**, *38*, 3518.
19. Yamanoi, Y. ; Imamoto, T. *J. Org. Chem.* **1999**, *64*, 2988.
20. Morimoto, T.; Yamaguchi, Y.; Suzuki, M.; Saitoh, A. *Tetrahedron Lett.* **2000**, *41*, 10025.
21. Degrado, S. J.; Mizutani, H.; Hoveyda, A. H. *J. Am. Chem. Soc.* **2001**, *123*, 755.
22. Alexakis, A.; Winn, C. L.; Guillen, F.; Pitkowicz, J.; Roland, S.; Mangeney, P. *Adv. Synth. Catal.* **2003**, *345*, 345.
23. a) Schinnerl, M.; Seitz, M.; Kaiser, A.; Reiser, O. *Org. Lett.* **2001**, *3*, 4259 and *Org. Lett.* **2002**, *4*,471.
24. Takemoto, Y.; Kuraoka, S.; Hamaue, N.; Aoe, K.; Hiramatsu, H.; Iwata, C. *Tetrahedron* **1996**, *52*, 14177.
25. Fraser, P. K.; Woodward, S. *Chem. Eur. J.* **2003**, *9*, 776.
26. Chataigner, I.; Gennari, C.; Ongeri, S.; Piarulli, U.; Ceccarelli, S. *Chem. Eur. J.* **2001**, *7*, 2628
27. Liang, L.; Chan, A. S. C. *Tetrahedron: Asymmetry* **2002**, *13*, 1393.
28. Kitamura, M.; Miki, T.; Nakano, K.; Noyori, R. *Bull. Chem. Soc. Jpn.* **2000**, *73*, 999.
29. Alexakis, A.; Benhaim, C.; Fournioux, X.; Van der Heuvel, A.; Levêque, J.-M.; March, S.; Rosset, S. *Synlett* **1999**, 1811.
30. Alexakis, A.; Benhaim, C.; Rosset, S.; Humam, M. *J. Am. Chem. Soc.* **2002**, *124*, 5262.
31. Knochel, P.; Singer, R.D. *Chem. Rev.* **1993**, 93, 2117.
32. Hayashi, T. *Synlett* **2001**, 879.
33. a) Escher, I. H.; Pfaltz, A. *Tetrahedron* **2000**, *56*, 2879. b) Arnold, L. A.; Naasz, R.; Minnaard, J.; Feringa, B. L. *J. Am. Chem. Soc.* **2001**, *123*, 5841.
34. Feringa, B. L. A*cc Chem. Res.* **2000**, *33*, 346.
35. Degrado, S. J.; Mizutani, H.; Hoveyda, A. H. *J. Am. Chem. Soc.* **2002**, *124*, 779.
36. a) Wendisch, V.; Sewald, N. *Tetrahedron: Asymmetry* **1997**, *8*, 1253. b) Versleijen, J. P. G.; Van Leusen, A. M.; Feringa, B. L. *Tetrahedron Lett.* **1999**, *40*, 5803. c) Alexakis, A.; Benhaim, C. *Org. Lett.* **2000**, *2*, 2579. d) Ongeri, S.; Piarulli, U.; Jackson, R. F. W.; Gennari, C. *Eur. J. Org. Chem.* **2001**, 803. e) Luchaco-Cullis, C. A. ; Hoveyda, A. H. *J. Am. Chem. Soc.* **2002**, *124*, 8192.
37. Alexakis, A.; Benhaim, C. *Tetrahedron: Asymmetry* **2001**, *12*, 1151.
38. Hird, A. W.; Hoveyda, A. H. *Angew. Chem., Int. Ed. Engl.* **2003**, *42*, 1276.

39. a) Villacorta, G. M.; Rao, C. P.; Lippard, S.J. *J. Am. Chem. Soc.* **1988**, 110, 3175-3182; b) Ahn, K. H.; Klassen, R. B.; Lippard, S. J. *Organomet.* **1990**, 9, 3178.
40. a) Lambert, F.; Knotter, D. M.; Janssen, M. D.; van Klaveren, M.; Boersma, J.; van Koten, G. *Tetrahedron : Asymmetry*, **1991**, *2*, 1097. b) Knotter, D. M.; Grove, D. M.; Smeets, W. J. J.; Speck, A. L.; van Koten, G. *J. Am. Chem. Soc.*, **1992**, *114*, 3400.
41. a) Zhou, Q. L.; Pfaltz, A. *Tetrahedron Lett.* **1993**, *34*, 7725. b) Zhou, Q. L.; Pfaltz, A. *Tetrahedron* **1994**, *50*, 4467.
42. Spescha, M.; Rihs, G. *Helv. Chim. Acta*, **1993**, *76*, 1219.
43. Seebach, D.; Jaeschke, G.; Pichota, A.; Audergon, L. *Helv. Chim. Acta.* **1997**, *80*,2515.
44. a) Kanai, M.; Nakagawa, Y.; Tomioka, K. *Tetrahedron* **1999**, *55*, 3843. b) Kanai, M.; Tomioka, K. *Tetrahedron Lett.* **1995**, *36*, 4275.
45. Stangeland, E.L.; Sammakia, T. *Tetrahedron* **1997**, *53*, 16503.
46. Keller, E.; Maurer, J.; Naasz, R.; Schader, T.; Meetsma, A.; Feringa, B. L. *Tetrahedron: Asymmetry* **1998**, *9*,2409.
47. Alexakis, A.; Trevitt, G. P.; Bernardinelli, G. *J. Am. Chem. Soc.* **2001**, *123*, 4358.
48. a) Naasz, R.; Arnold, L. A.; Pineschi, M.; Keller, E.; Feringa, B. L. *J. Am. Chem. Soc.* **2001**, *123*, 4358. b) Naasz, R.; Arnold, L. A.; Minnaard, A. J.; Feringa, B. L. *Chem. Comm.* **2001**, 735.
49. Degrado, S. J.; Mizutani, H.; Hoveyda, A. H. *J. Am. Chem. Soc.* **2002**, *124*, 13362.
50. Alexakis, A.; Knopff, O. *Org. Lett.* **2002**, *4*,3835.
51. Alexakis, A.; March, S. *J. Org. Chem.* **2002**, *67*, 8753.
52. Breit, B.; Demel, P. in *Modern organocopper chemistry*, Krause, N; Ed., Wiley-VCH, Weinheim, **2002**, 188.
53. Yamamoto Y. *Angew. Chem. Int. Ed. Engl.* **1986**, *25*, 947.
54. (a) Gallina, C.; Ciattini, P.G. *J. Am. Chem. Soc.*, **1979**, *101*, 1035. (b) Goering, H. L.; Kantner, S. S.; Tseng, C. C. *J. Org. Chem.* **1983**, *48*, 715.
55. Breit, B.; Demel, P. *Adv. Synth. Catal.* **2001**, *343*, 429.
56. Alexakis, A.; Mangeney, P.; Ghribi, A.; Marek, I.; Sedrani, R.; Guir, C.; Normant J. F. *Pure Appl. Chem.* **1988**, *60*, 49.
57. Denmark, S. E.; Marble, L. K. *J. Org. Chem.* **1990**, *55*, 1984.
58. Calo, V.; Fiandese, V.; Nacci, A.; Scilimati, A. *Tetrahedron* **1994**, *50*, 7283.
59. Gais, H.-J.; Müller, H.; Bund, J.; Scommoda, M.; Brandt, J.; Raabe, G. *J. Am. Chem. Soc.*, **1995**, *117*, 2453.
60. a) van Klaveren, M.; Personn, E. S. M.; del Villar, A.; Grove, D. M.; Bäckvall, J.-E.; van Koten, G. *Tetrahedron Lett.*, **1995**, *36*, 3059. b)

Karltröm, A. S. E.; Huerta, F. F.; Meuzelaar, G. J.; Bäckvall, J.-E *Synlett* **2001,** 923.

61. a) Alexakis, A.; Malan, C.; Lea, L.; Benhaim, C.; Fournioux, X. *Synlett* **2001,** 927. b) Alexakis, A.; Croset, K *Org. Lett.* **2002,** *4*,4147.
62. a) Dübner, F.; Knochel, P. *Angew. Chem. Int. Ed. Engl.* **1999,** *38*, 379. b) Dübner, F.; Knochel, P. *Tetrahedron Lett.* **2000,** *41*, 9233.
63. Malda, H.; van Zijl, A.W.; Arnold, L.A.; Feringa, B. L. *Org. Lett.* **2001,** *3,* 1169.
64. Luchaco-Cullis, C. A.; Mizutani, H.; Murphy, K. E.; Hoveyda, A. H. *Angew. Chem. Int. Ed.* **2001**, *40*, 1456.
65. Borner, C.; Goldsmith, P. J.; Woodward, S.; Gimeno, J.; Gladiali, S.; Ramazzotti, D. *Chem. Commun.* **2000**,2433.
66. a) Ongeri, S.; Piarulli, U.; Roux, M.; Monti, C.; Gennari, C. *Helv. Chim. Acta* **2002**, *85*, 3388. b) Piarulli, U.; Daubos, P.; Claverie, C.; Roux, M.; Gennari, C. *Angew. Chem. Int. Ed.* **2003**, *42*, 234.
67. Lipshutz, B. H.; Sengupta S. *Org. Rect.* **1992**, *41*, 135.
68. Davies, S.G., Wollowitz *Tetrahedron Lett.*, **1980**, *21*, 4175.
69. a) Müller, P.; Nury P. *Org. Lett.* **1999**, *1*, 439. b) Müller, P.; Nury P. *Helv. Chim. Acta* **2001**, *84*, 662.
70. Badalassi, F.; Crotti, P.; Macchia, F.; Pineschi, M.; Arnold, A.; Feringa, B. L. *Tetrahedron Lett.*, **1998**, *39*, 7795.
71. a) Bertozzi, F.; Crotti, P.; Macchia, F.; Pineschi, M.; Arnold, A.; Feringa, B. L. *Tetrahedron Lett.*, **1999**, *40*, 4893. b) Bertozzi, F.; Crotti, P.; Macchia, F.; Pineschi, M.; Feringa, B. L. *Angew. Chem. Int. Ed.* **2001**, *40*, 930. c) Bertozzi, F.; Pineschi, M.; Macchia, F.; Arnold, L. A.; Minnaard, A. J.; Feringa, B. L. *Org. Lett.* **2002**, *4*, 2703.

Chapter 5

Chiral Imidazolylidine Ligands for Asymmetric Catalysis

Kevin Burgess[1], Marc C. Perry[1,2], and Xiuhua Cui[1]

[1]Department of Chemistry, Texas A&M University, College Station, TX 77843–3255
[2]Current address: Department of Chemistry, University of Alaska Anchorage, 3211 Providence Drive, Anchorage, AK 99508

Combination of a small library of oxazoline electrophiles and imidazole nucleophiles led to the formation of a significant number of chiral imidazolium salts. Iridium and rhodium complexes were synthesized from these by the treatment with lithium *tert*-butoxide in the presence of $[Ir(COD)Cl]_2$ or $[Rh(NBD)_2][BF_4]$. These complexes were screened in the hydrogenation of unfunctionalized alkenes.

The use of *N*-heterocyclic carbenes as ligands in transition metal catalysis has increased rapidly since the discovery of the first stable imidizolylidene by Arduengo *et al* in the early 1990's. Use of such ligands has led to significant advances in non-enantioselective catalysis particularly for ruthenium-catalyzed double bond metathesis(*1,2*) and palladium-catalyzed C-C bond formation.(*3*) Chiral analogues of these *N*-heterocyclic carbenes have been developed,(*4-6*) but few have resulted in high enantioselectivities in any reaction, and the ones that have given high enantioselectivities have only been tested on a narrow class of substrates. A partial list of the chiral *N*-heterocyclic carbenes which have been successfully used in asymmetric transformations is shown in Figure 1.
We synthesized a collection of a new class of imidazol-2-ylidine-oxazoline ligand precursors **1** by combination of a small library of oxazoline electrophiles **2** with a small library of monosubstituted imidazoles **3** as shown in Figure 2.

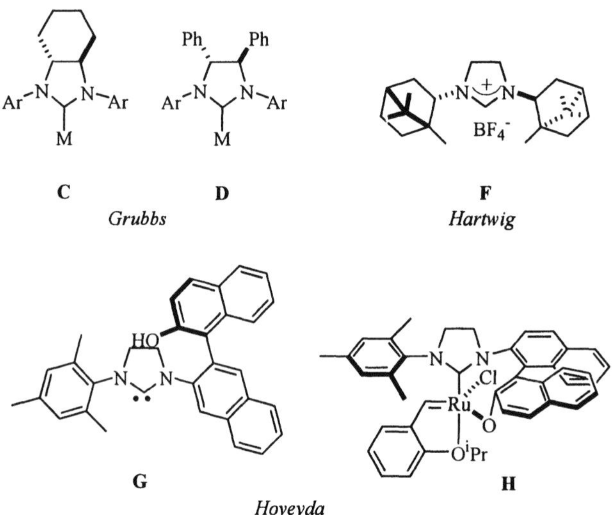

Figure 1. Some of the successful chiral N-heterocyclic carbene ligands and complexes which have been prepared.

3	2	1
small library	*small library*	*large library*

Figure 2. Synthesis of a library of imidazol-2-ylidene-oxazoline ligands.

This approach has significant advantages because a relatively large library of ligands can be made from small numbers of components. This library of ligands can then be screened in a variety of reactions using multiple substrates. Here we describe use of this approach for the synthesis of imidazol-2-ylidine-oxazoline iridium(I) complexes **4** and the screening of these complexes in the asymmetric hydrogenation of unfunctionalized aryl alkenes.(*7*) Similar rhodium(I) complexes **5** were also synthesized and screened.

Library Synthesis

Synthesis of Oxazoline Electrophiles

In previous studies we had developed a synthesis of a series of *P,N*-ligands that we have called JM-Phos.(*8*) In that work several routes to oxazoline tosylates were devised, the most general of which was used to synthesize the six oxazoline iodides **2a-f** as shown in Figure 3. These were competent electrophiles for the coupling with monosubstituted imidazoles.

Synthesis of Monosubstituted Imidazoles

A library of seventeen monosubstituted imidazoles **3a-q** was prepared using three different synthetic methods. The first, shown in Figure 4, is a one-pot condensation that works well when aliphatic amines are used.(*9*) The second method, shown in Figure 5, is a Cu-catalyzed coupling that works well for unhindered aryl bromides.(*10*) The last route, shown in Figure 6, is a multistep procedure which allows for the formation of sterically demanding imidazoles starting from aromatic amines.(*11*) This final method works quite well on relatively large scales.

Synthesis of Imidazolium Salts as Ligand Precursors

A library of imidazolium salts **1** was prepared by reacting imidazole nucleophiles with oxazoline electrophiles as shown in Figure 7. The reaction is carried out in DMF at 100 °C over 12 h. The salt was isolated by cooling the DMF solvent and precipitating with ether. After filtration, the salt was used in the synthesis of iridium and rhodium complexes without further purification. These ligand precursors are completely air stable giving them a significant advantage over conventional phosphine analogues.

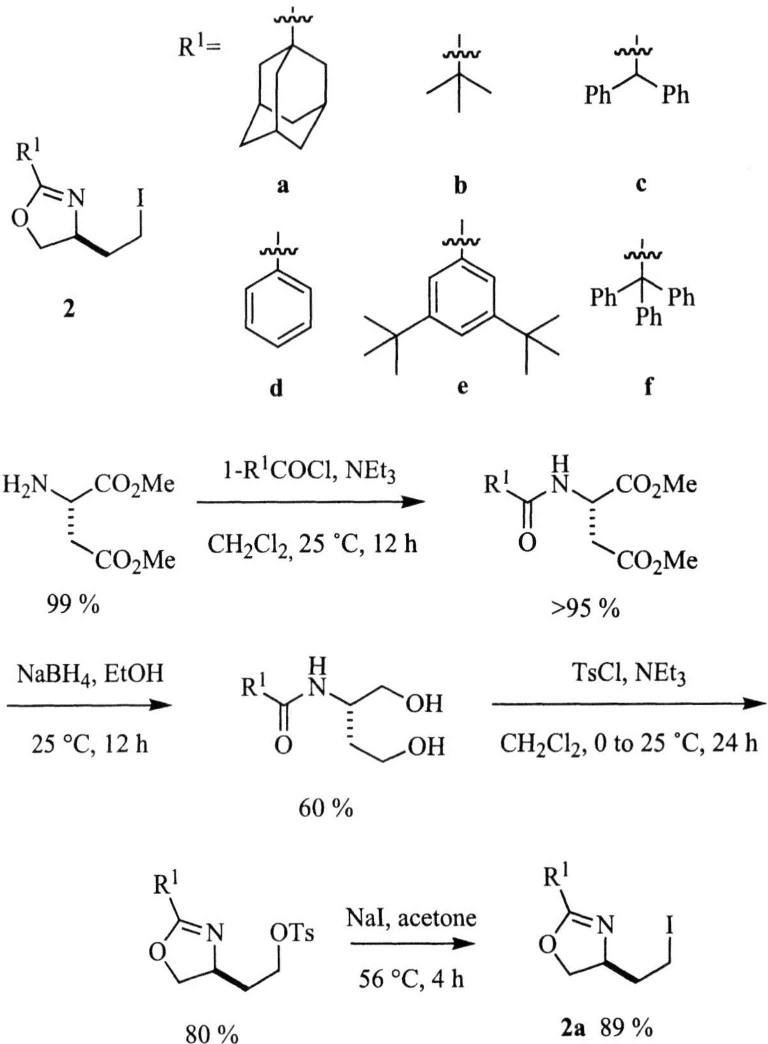

Figure 3. Synthesis of a library of oxazoline iodides

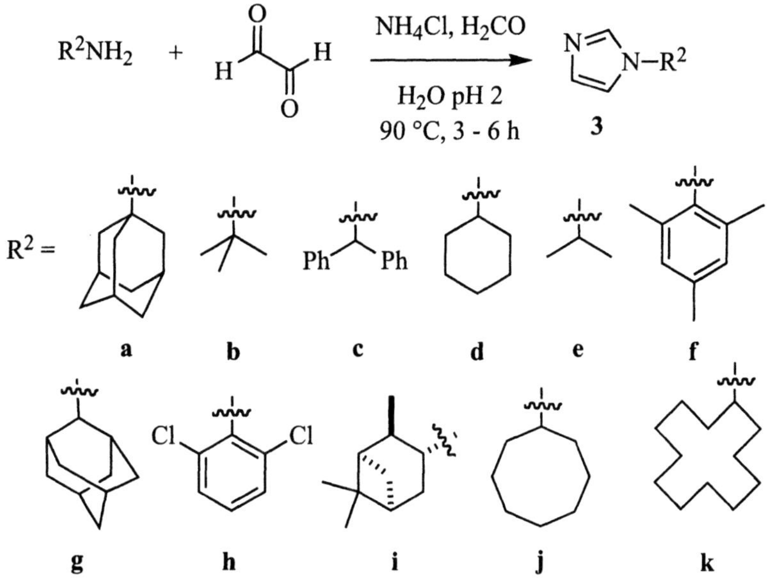

Figure 4. Synthesis of monosubstituted imidazoles by a one-pot condensation.

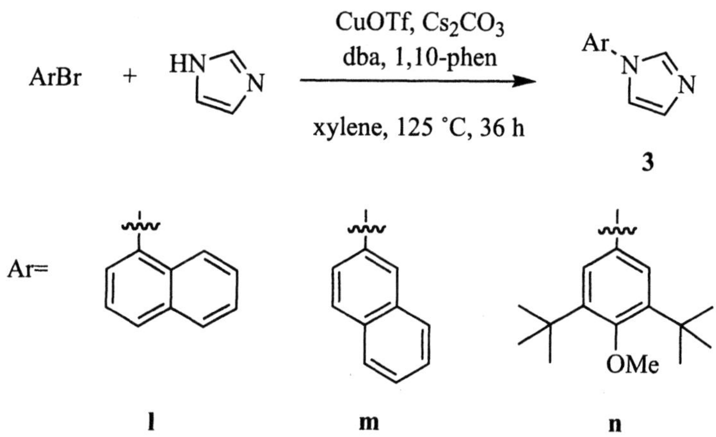

Figure 5. Copper-catalyzed synthesis of monosubstituted imidazoles.

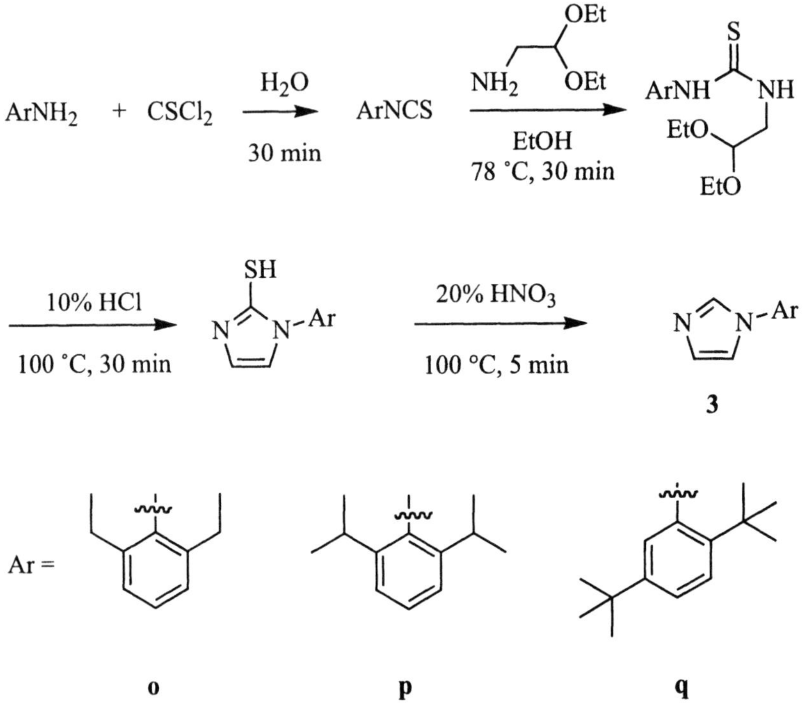

$ArNH_2$ + $CSCl_2$ →(H_2O, 30 min) $ArNCS$ →($NH_2CH_2CH(OEt)_2$, EtOH, 78 °C, 30 min) ArNH–C(=S)–NH–$CH_2CH(OEt)_2$

→(10% HCl, 100 °C, 30 min) 2-mercapto-1-Ar-imidazole →(20% HNO_3, 100 °C, 5 min) 1-Ar-imidazole **3**

Ar = **o**, **p**, **q**

Figure 6. Multistep synthesis of hindered N-aryl imidazoles.

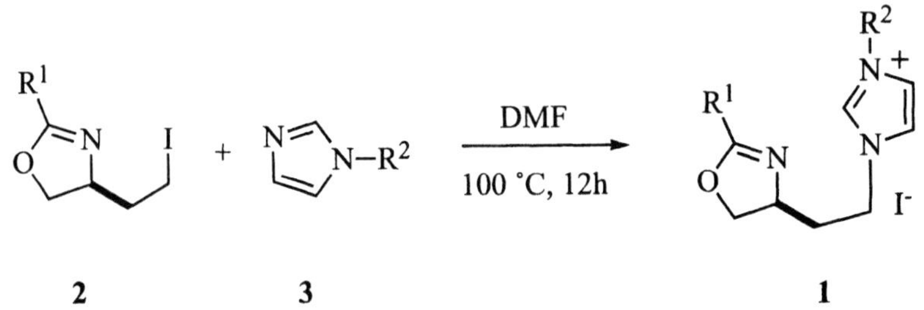

2 + **3** →(DMF, 100 °C, 12h) **1** (I^-)

Figure 7. Synthesis of a library of imidazolium salts as ligand precursors.

Synthesis of Iridium Complexes

A library of iridium complexes **4** was prepared as shown in Table I. The imidazolium salt was deprotonated with lithium *tert*-butoxide forming the carbene ligand which reacted with the $[Ir(COD)Cl]_2$ present in solution. The iridium complexes were isolated by column chromatography after anion exchange with NaBARF. They were found to be air stable, easily manipulated materials. However, in a few cases, where both the oxazoline substituent and imidazole substituent were very bulky, the complex formation was not clean and the desired complex was not isolated (e.g. for R^1 = 3,5-$^tBu_2C_6H_3$; R^2 = 2,6-$^iPr_2C_6H_3$). Combination of the library of 6 oxazoline iodides with 17 monosubstituted imidazoles had the potential to produce a library of 102 iridium complexes. The number that we synthesized was significantly smaller because screening began as soon as the first batches of ligands were prepared and the results allowed us to focus our library.

Table I. Iridium Complexes

(i) $[Ir(COD)Cl]_2$, LiO^tBu
THF, 70 °C, 16 h

(ii) NaBARF
1:1 $H_2O:CH_2Cl_2$

$BARF^-$

4

4	R^1	R^2
a	1-Ad	^tBu
b	1-Ad	$CHPh_2$
c	1-Ad	Cy
d	1-Ad	2,4,6-$Me_3C_6H_2$
e	1-Ad	3,5-$^tBu_2$-4-$MeOC_6H_2$
f	1-Ad	2,6-$Et_2C_6H_3$
g	1-Ad	2,6-$^iPr_2C_6H_3$
h	1-Ad	2,5-$^tBu_2C_6H_3$
i	^tBu	1-Ad
j	^tBu	^tBu
k	^tBu	$CHPh_2$
l	^tBu	2,6-$^iPr_2C_6H_3$
m	$CHPh_2$	2,6-$^iPr_2C_6H_3$
n	Ph	2,6-$^iPr_2C_6H_3$

Synthesis of Rhodium Complexes

A small library of rhodium complexes **5** was synthesized from $[Rh(NBD)_2][BF_4]$ in a similar manner to the iridium complexes as shown in Figure 8. Far fewer rhodium complexes were synthesized because they were found to give poor chemical and optical yields in the hydrogenation of unfunctionalized alkenes.

Screening

Initial Screening

The very first complexes prepared contained the 2,6-di*iso*propylphenyl group attached to the carbene portion of the ligand, and differed only in the oxazoline substituent. These complexes were screened in the hydrogenation of *E*-α-methylstilbene (Figure 9). This substrate was chosen because others found it to be the easiest to hydrogenate with high enantioselectivities.(*12*) It was found that only the adamantyl substituent on the oxazoline gave satisfactory yields and enatioselectivities. Drastic reactivity differences between ligands with adamantyl and *tert*-butyl substituents seemed surprising, but was found to be true in hydrogenation of other stilbene derivatives as well (Figure 10). Thus only slight changes in ligand structure for these reactions had marked effects on catalyst performance.

Other data obtained in the course of this study underline the subtle interplay of catalyst performance and structure. Thus changing the di*iso*propylphenyl group into a diethylphenyl group on the imidazole part of the ligand, while retaining the adamantyl substituent on the oxazoline, had a significant effect on catalyst performance for the two substrates shown in Figure 11.

The particular case of ligands from *N*-alkyl imidazoles gave a particulary interesting, but negative, result. Virtually no yield of product was obtained in the hydrogenation of α-methylstilbene, as shown in Figure 12, using such ligands, even though the corresponding *N*-aryl systems were highly active. In fact, the initial complex could be recovered unchanged after the reaction; the COD ligand had not been removed.(*13,14*) Further, when complex **4c** and **4g** were treated individually with 1 atm of hydrogen in $CDCl_3$ for 3 h, **4g** was converted cleanly into a new species while complex **4c** remained unchanged.

To determine if the difference in reactivity between the *N*-alkyl and *N*-aryl imidazolylidine ligands had electronic origins, we synthesized the carbonyl complexes **6** and **7** by stirring **4g** and **4c** respectively under an atmosphere of CO for 3 h. The IR stretches of the carbonyls were identical indicative of similar electronic back-donation effects from the metal. We therefore conclude that the

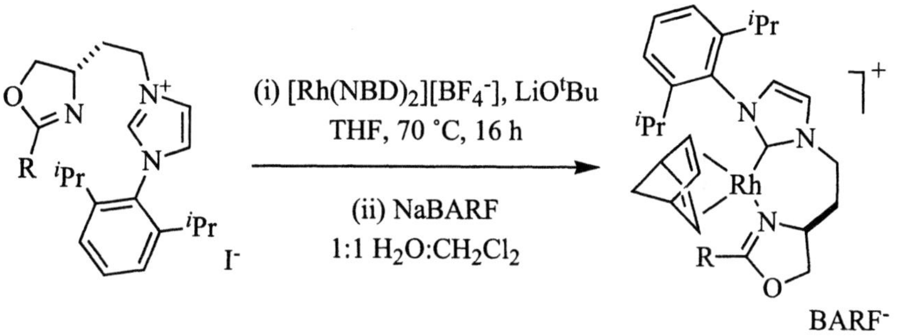

(i) $[Rh(NBD)_2][BF_4^-]$, LiO^{t}Bu
THF, 70 °C, 16 h

(ii) NaBARF
1:1 $H_2O:CH_2Cl_2$

1

5

R = Ph, **a**, 47 %;
^{t}Bu, **b**, 43 %

Figure 8. Synthesis of rhodium complexes

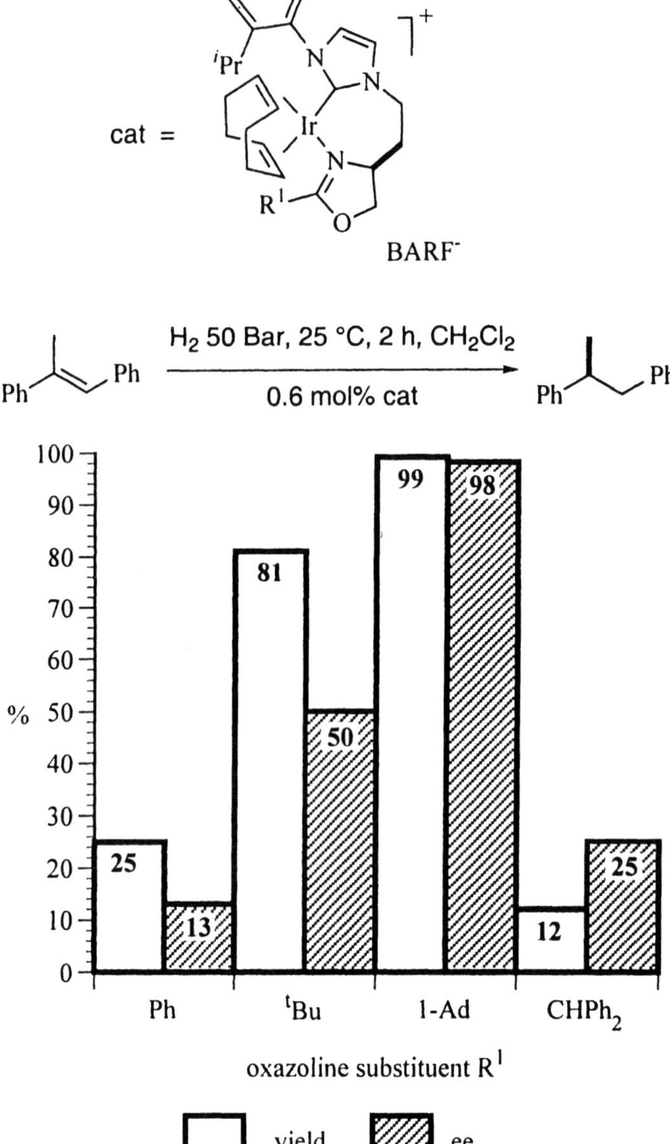

Figure 9. Initial screen for hydrogenation of methylstilbene

cat =

^{i}Pr N N Ir N R^1 O ^{i}Pr]$^+$

BARF$^-$

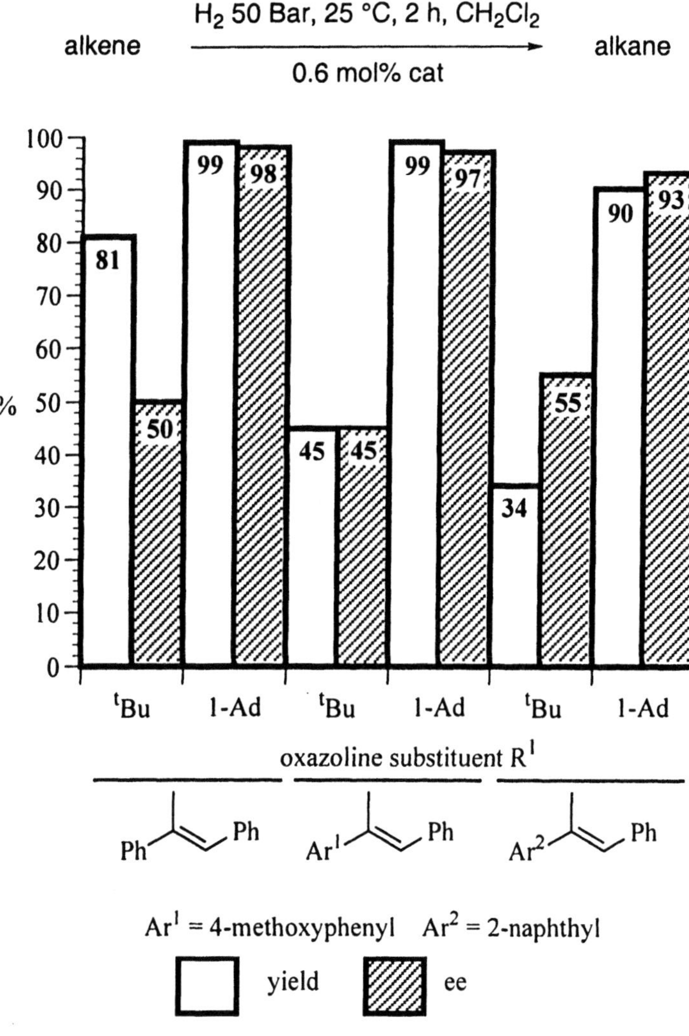

Figure 10. Hydrogenation of 2-arylpropenes

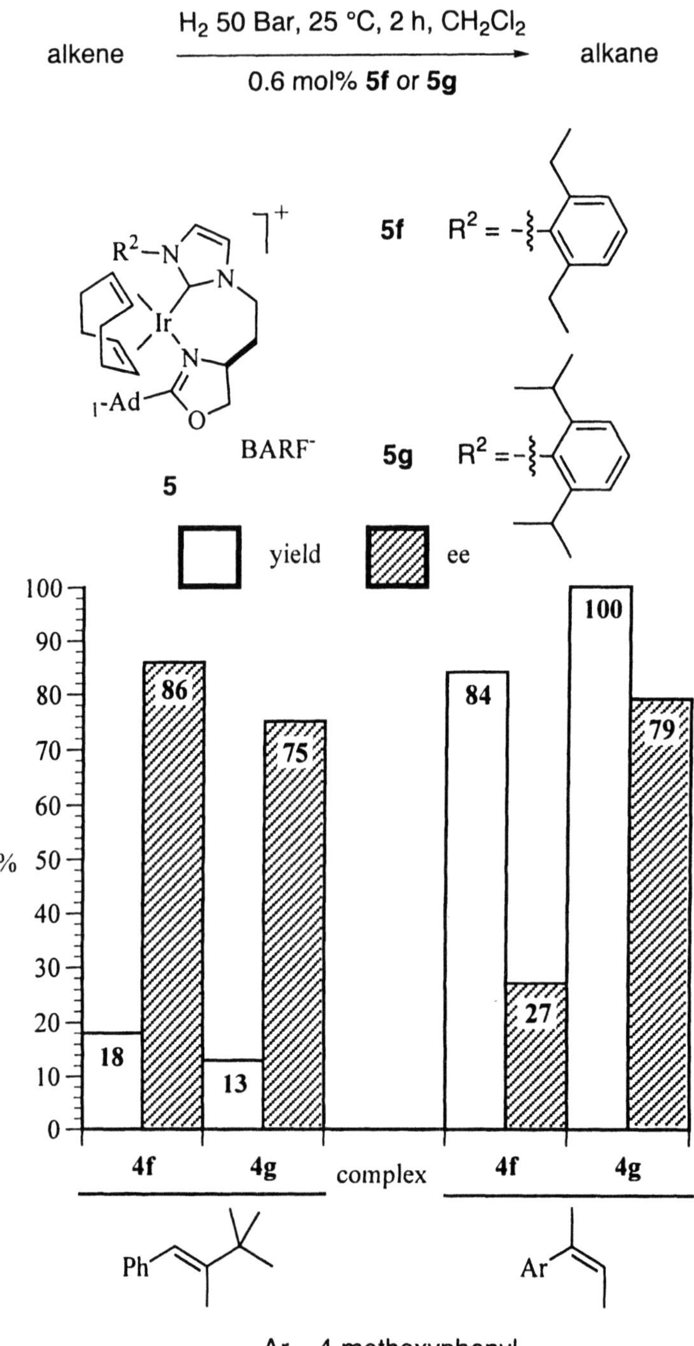

Ar = 4-methoxyphenyl

Figure 11. Comparison of catalysts with different imidazole substituents

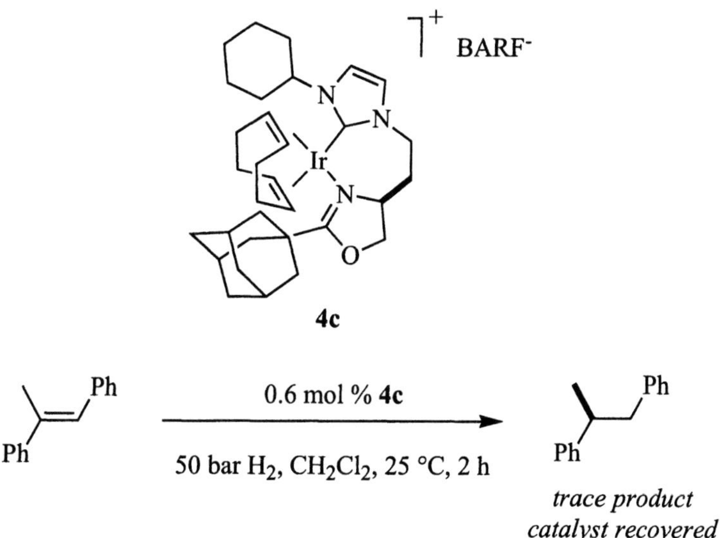

Figure 12. Hydrogenation using a catalyst with an N-alkyl imidazole substituent.

reactivity differences are probably due to steric rather than electronic differences.

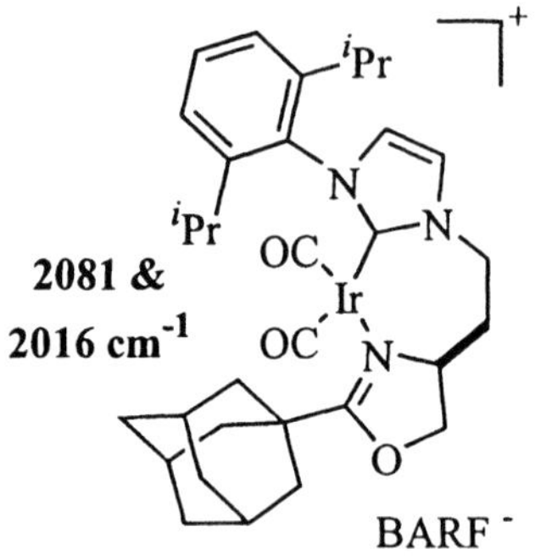

2081 & 2016 cm^{-1} **2082 & 2015 cm^{-1}**

6 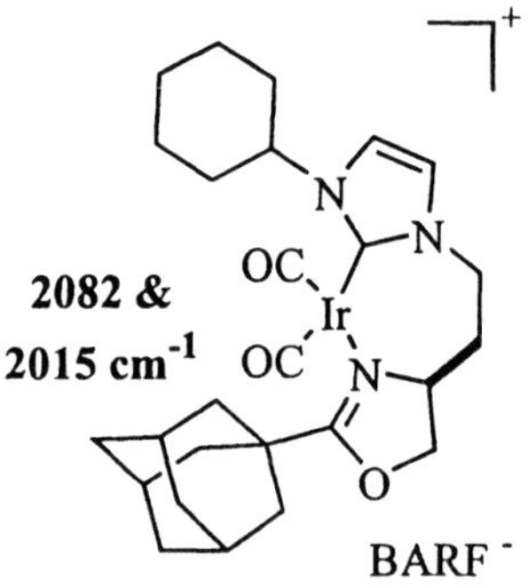 **7**

After the initial screening, complex **4g** emerged as the best catalyst in many situations. If the key structural features responsible for its reactivity and high enantiodiscrimination could be identified this would be valuable for the design of other catalysts. We were able to obtain X-ray quality crystals of complexes **4a** and **4c**, but complex **4g** failed to crystallize under any of the conditions tried. However, the corresponding rhodium complex **5c** of the same ligand was found to crystallize nicely. The structures of these complexes were solved and are represented in Figure 13 (with the COD ligand removed for clarity). In complex **5g**, the di*iso*propylphenyl group projects into the bottom left quadrant, while the cyclohexyl and *tert*-butyl in complexes **4c** and **4a** do not. We believe that this may explain why complexes of this ligand give higher enantioselectivities than ones which do not have any steric presence in this region of the metal center.

Screening of Additional Substrates

Complex **4g** was screened in hydrogenations of other substrates and the results are shown in Figure 14. *E*-Alkenes gave alkanes with higher optical activities than the corresponding *Z*-alkenes. At the time this work was completed, the yields and enantioselectivities obtained using **4g** were comparable with the best results in the literature reported by others using *P,N*-ligands. Screening of other complexes in our library failed to give better results than complex **4g** for any of these substrates.

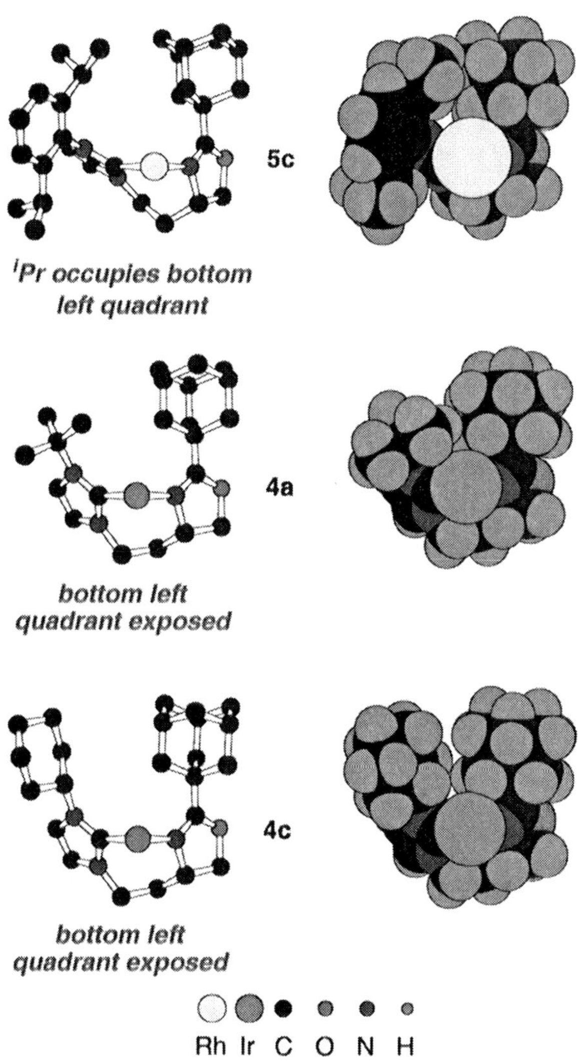

*Figure 13. Partial structures of **5c, 4a** and **4c**. (See page 1 of color insert.)*

$$\text{alkene} \xrightarrow[\text{0.6 mol\% } \mathbf{5g}]{H_2 \text{ 50 Bar, 25 °C, 2 h, } CH_2Cl_2} \text{alkane}$$

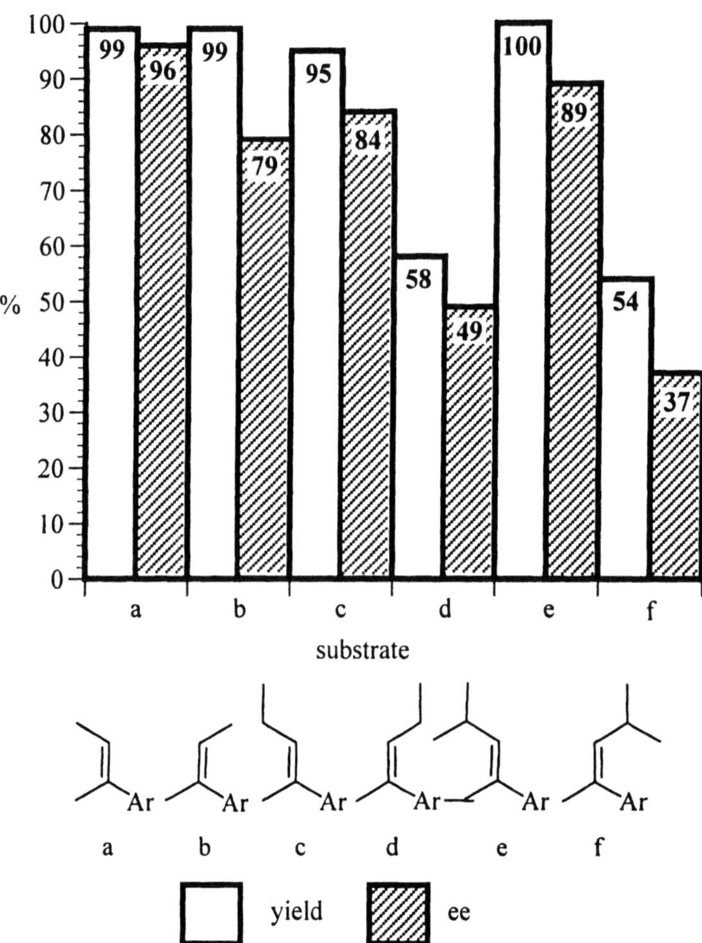

Figure 14. Hydrogenation of multiple substrates using catalyst ***4g.***

Effects of Reaction Parameters

Temperature and Pressure Effects

When we started work in this area, temperature had been shown to effect enantioselectivities and yields obtained using *P,N*-ligands, but only to a limited extent, and pressure was shown to have no effect on yield and *ee* using *E*-α-methylstilbene as a substrate.(*12,15,16*) No data on the pressure effects for other substrates had been reported and all the reactions were run at or above 50 bar of hydrogen pressure. We hoped to find a catalyst that was effective at atmospheric pressure.

The effect of temperature was investigated using three substrates as shown in Figure 15. Temperature changes had little effect on the yield or enantioselectivities obtained for either *E*- or *Z*-2-phenyl-2-butene over the range 0 - 40 °C. However, temperature had a dramatic effect on enantioselectivity for the hydrogenation of 2-phenyl-1-butene; it varied between 26 % *ee* of the *S*-isomer at 0 °C and 69% *ee* of the *R*-isomer at 40 °C.

Pressure effects were also investigated using the 5 substrates shown in Figure 16. There was no significant effect on the yields or enantioselectivities for three of the substrates, but profound effects were observed for the other two. All the substrates gave optimal yields and enatioselectivities at 1 atm of hydrogen. This is the first example of a catalyst which is shown to be capable of hydrogenating a variety of alkenes in good yields and enantioselectivities at atmospheric pressure.

One of the substrates which was effected by pressure was 2-phenyl-1-butene, the same substrate shown to exhibit a dramatic temperature effect. By looking at the temperature and pressure effects for this substrate it is likely that the concentration of hydrogen in solution is actually the variable which determines the enantioselectivity for this substrate.

Optimization of Conditions for Hydrogenation of 2-Phenyl-1-butene

Conditions were changed to see what the optimal enantioselectivies that could be obtained in either direction for the hydrogenation of 2-phenyl-1-butene. The enantioselectivies ranged from 64 % of the *S*-isomer at high hydrogen concentrations (85 bar H_2, -15 °C) to 89 % of the *R*-isomer at low hydrogen concentrations (1 bar H_2, 30 °C) as shown in Figure 17. This dramatic change in enantioselectivity indicates a change in the predominant mechanism between these two sets of conditions.

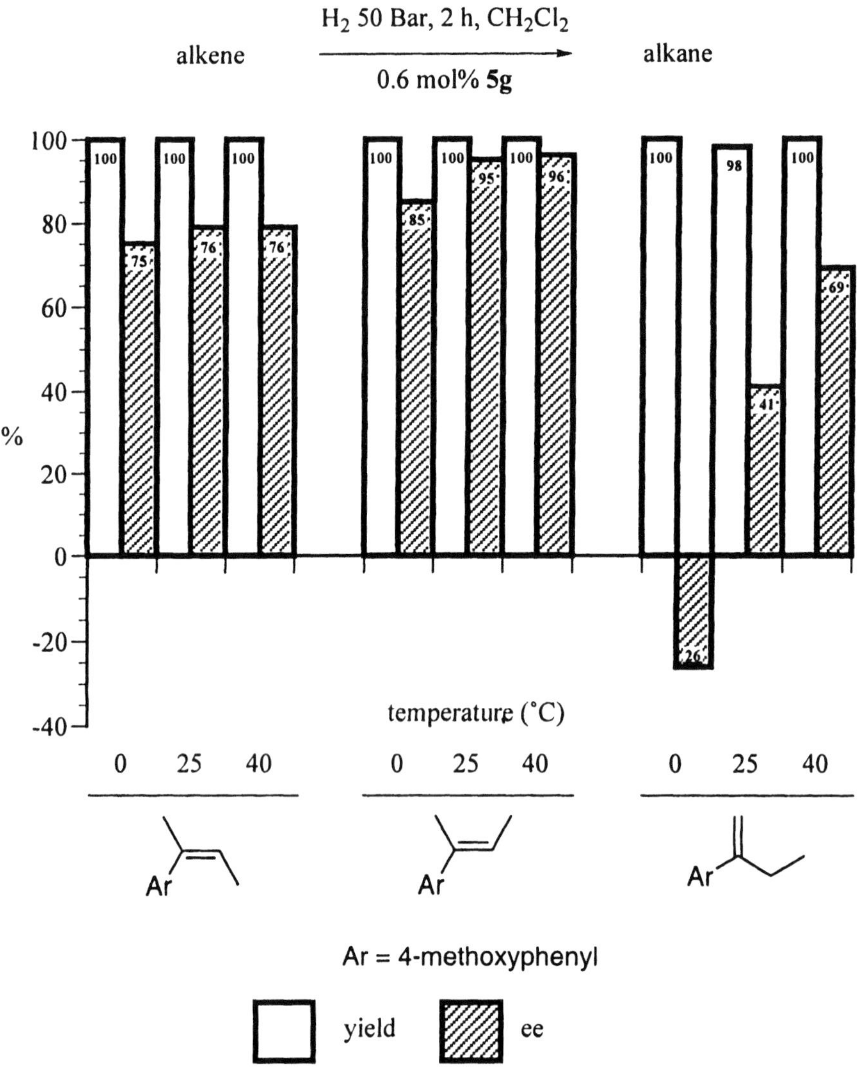

Figure 15. Temperature effects

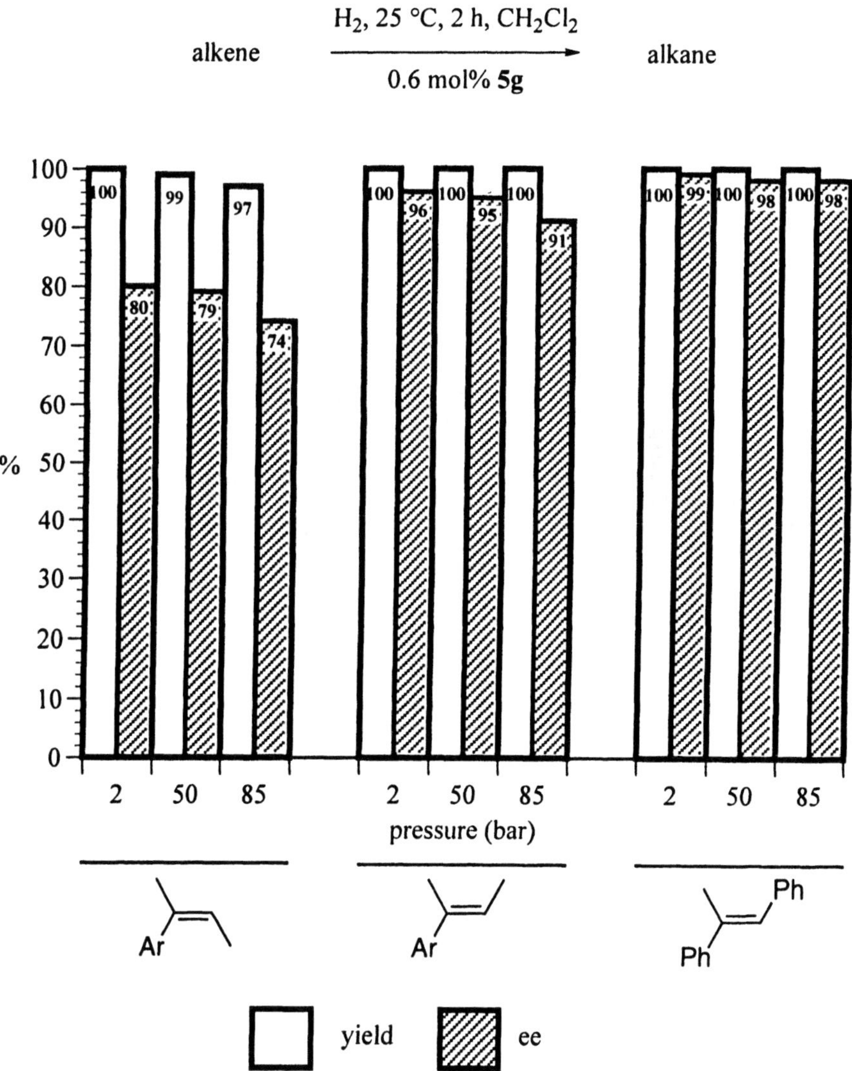

Figure 16. Pressure effects. *Continued on next page.*

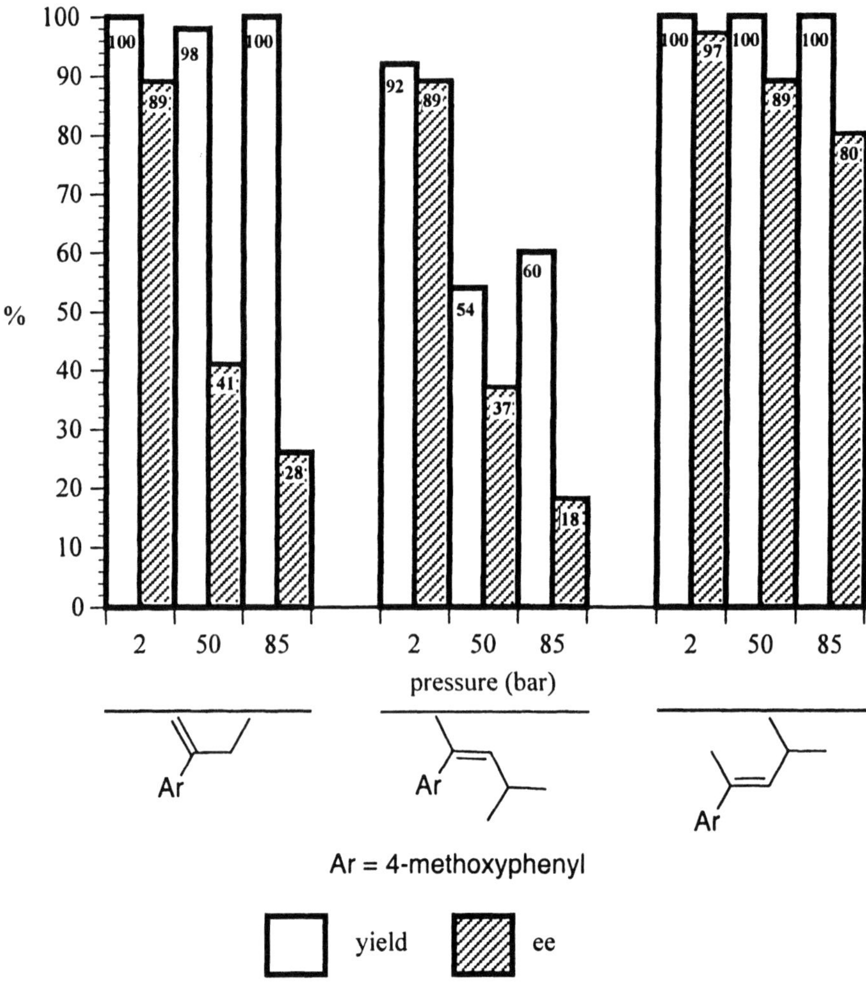

Figure 16. Continued.

Preliminary Mechanistic investigation

Deuteration Experiments

One possible explanation for the difference in enantioselectivities at varying hydrogen concentrations is that under certain conditions double bond migration occurs prior to hydrogenation while under different conditions direct hydrogenation predominates. This has been demonstrated in Ru-catalyzed hydrogenation of geraniol.(*17*) To investigate this possibility we performed deuteration under the conditions which gave maximum enantioselectivities of both enantiomers as shown below. Both the high pressure/low temperature and low pressure/high temperature conditions had some deuterium incorporation on a carbon not originally involved in the double bond, but the incorporation was not high enough to indicate double bond migration prior to deuteration.

Ar
0.2 M
0.1 mol% **4g**
1 bar D_2, CH_2Cl_2
23 °C, 2 h
(1.00) D
D (0.88)
Ar
D (0.11)

Ar
0.2 M
0.6 mol% **4g**
75 bar D_2, CH_2Cl_2
-15 °C, 2 h
(0.99) D
D (1.00)
Ar
D (0.08)

Ar
0.2 M
0.6 mol% **4g**
1 bar D_2, CH_2Cl_2
23 °C, 2 h
(0.23) D
D (0.79)
Ar
D (1.00)

Ar
0.2 M
0.6 mol% **4g**
50 bar D_2, CH_2Cl_2
23 °C, 2 h
(0.13) D
D (0.91)
Ar
D (1.00)

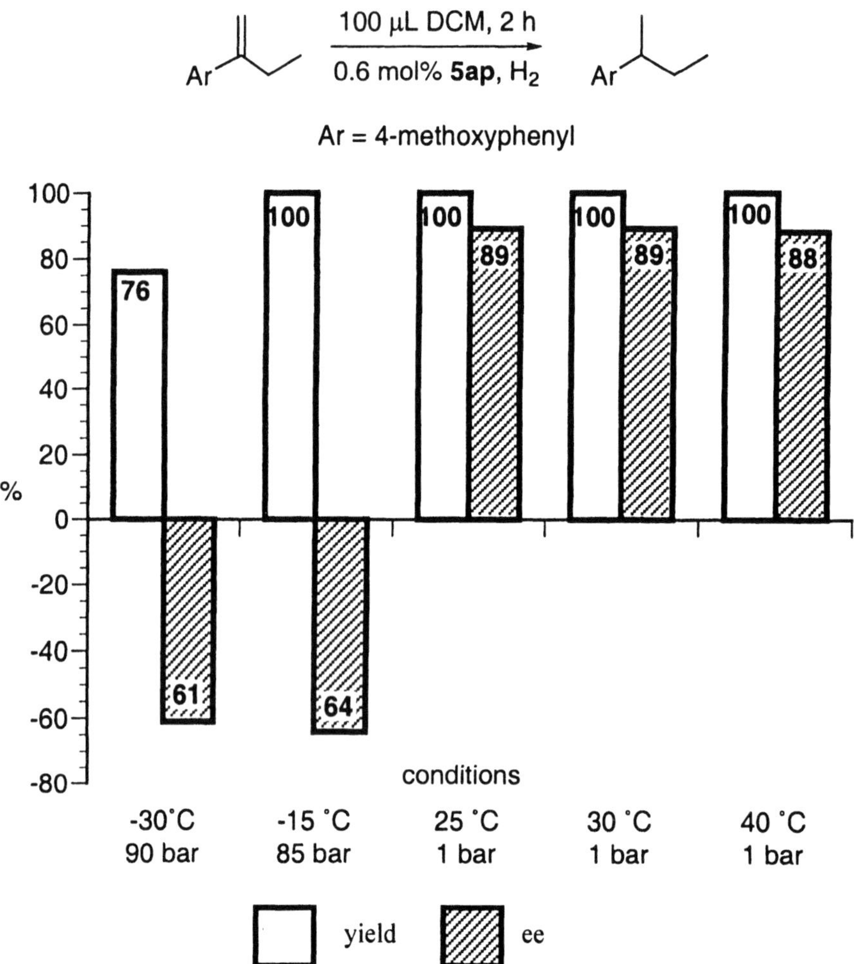

Figure 17. Optimization of hydrogenation conditions for 2-phenyl-2-butene

Conclusions

A library of new imidazol-2-ylidine-oxazoline ligands was synthesized and both iridium and rhodium complexes of these were easily prepared. The complexes could be purified by column chromatography and were found to be air stable. All of the rhodium complexes prepared failed to produce acceptable chemical and optical yields in the hydrogenation of arylalkenes. The iridium complexes were more promising, and one of the iridium complexes obtained, **4g**, was found to be competitive with the best catalysts known for the hydrogenation of unfunctionalized alkenes. Complex **4g**, unlike the complexes reported previously, gave optimal results at room temperature and atmospheric pressure of hydrogen. Analysis of the crystal structures of several complexes indicated that **4g** occupied a larger volume of space around the metal center compared to the others leaving a smaller chiral pocket. It would have been impossible to predict the results obtained beforehand, and the use of libraries proved invaluable in this case.

A variety of substrate classes were hydrogenated in good to excellent yields and enantioselectivities at ambient temperature and atmospheric pressures of hydrogen. *E*-alkenes were shown to be hydrogenated with higher enantioselectivities than the corresponding *Z*-alkenes. Temperature and pressure were found to have little effect for some substrates and profound effects for others.

The hydrogenation of 2-phenyl-1-butene was shown to give 64% ee of the *R*-isomer under high pressure/low temperature conditions and 89% ee of the *S*-isomer under low pressure/high temperature conditions. This result indicates that the predominant mechanism changes as the concentration of hydrogen in solution changes from high to low. Deuteration experiments indicate that this is not due to double bond migration prior to hydrogenation under one set of conditions.

The ligands prepared in this work were found to be effective for the hydrogenation of aryl alkenes when bound to iridium. Investigation into the use of these ligands in other metal-catalyzed processes is ongoing.

Acknowledgements.

Financial support for this work was provided by The Robert Welch Foundation and Johnson Matthey plc. We thank Dr Thomas A. Colacot (JM) for help and advice.

References

1. Trnka, T. M.; Grubbs, R. H. *Acc. Chem. Res.* **2001**, *34*, 18-29.
2. Jafarpour, L.; Nolan, S. P. *J. Organomet. Chem.* **2001**, *617-8*, 17-27.
3. Hillier, A. C.; Grasa, G. A.; Viciu, M. S.; Lee, H. M.; Yang, C.; Nolan, S. P. *J. Organomet. Chem* **2002**, *653*, 69-82.
4. Perry, M. C.; Burgess, K. *Tetrahedron Asymm.* **2003**, 951-961.
5. Seiders, T. J.; Ward, D. W.; Grubbs, R. H. *Org. Lett.* **2001**, *3*, 3225-3228.
6. Van Veldhuizen, J. J.; Garber, S. B.; Kingsbury, J. S.; Hoveyda, A. H. *J. Am. Chem. Soc.* **2002**, *124*, 4954-4955.
7. Perry, M. C.; Cui, X.; Powell, M. T.; Hou, D.-R.; Reibenspies, J. H.; Burgess, K. *J. Am. Chem. Soc.* **2003**, *125*, 113-123.
8. Hou, D.-R.; Reibenspies, J.; Burgess, K. *J. Org. Chem.* **2001**, *66*, 206-215.
9. Gridnev, A. A.; Mihaltseva, I. M. *Synth. Commun.* **1994**, *24*, 1547-1555.
10. Kiyomori, A.; Marcoux, J.-F.; Buchwald, S. L. *Tetrahedron Lett.* **1999**, *40*, 2657-2660.
11. Johnson, A. L.; Kauer, J. C.; Hsarma, D. C.; Dorfman, R. I. *J. Med. Chem.* **1969**, *12*, 1024-1028.
12. Blackmond, D. G.; Lightfoot, A.; Pfaltz, A.; Rosner, T.; Schnider, P.; Zimmermann, N. *Chirality* **2000**, *12*, 442-449.
13. Borner, A.; Heller, D. *Tetrahedron Lett.* **2001**, *42*, 223-225.
14. Drexler, H.-J.; Baumann, W.; Spannenberg, A.; Fischer, C.; Heller, D. *J. Organomet. Chem.* **2001**, *621*, 89-102.
15. Menges, F.; Pfaltz, A. *Adv. Synth. Catal.* **2002**, *344*, 40-44.
16. Blankenstein, J.; Pfaltz, A. *Angew. Chem. Int. Ed.* **2001**, *40*, 4445-4447.
17. Sun, Y.; Wang, J.; LeBlond, C.; Reamer, R. A.; Laquidara, J.; Sowa-Jr., J. R.; Blackmond, D. G. *J. Organomet. Chem.* **1997**, *548*, 65-72.

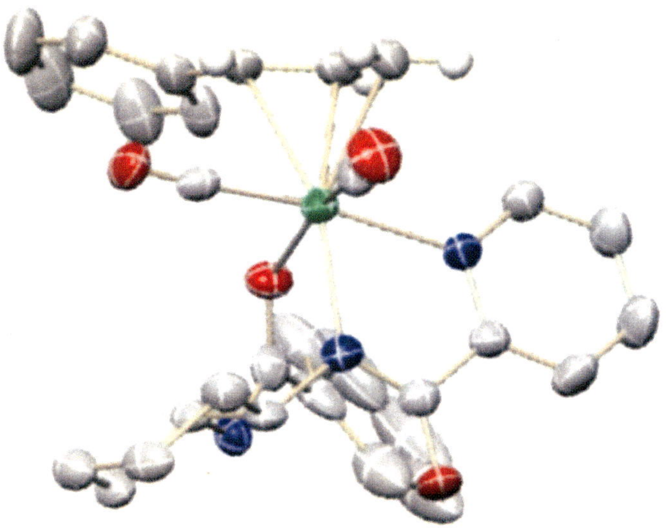

Figure 1. ORTEP plot of the crystal structure of complex **19**. Color coding: Molybdenum is green, carbon gray, hydrogen white, nitrogen blue, and oxygen red.

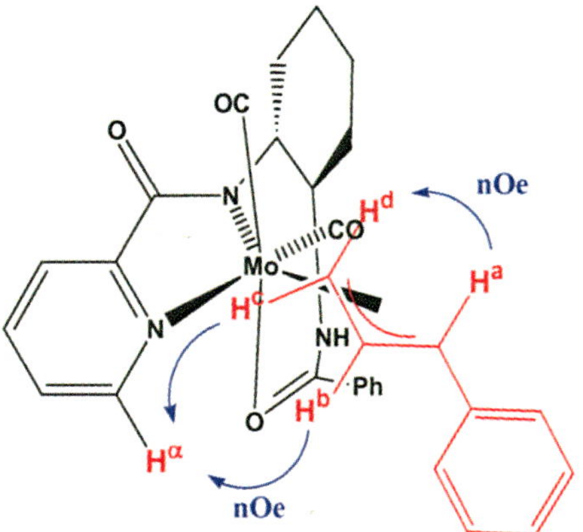

Figure 2. Solution structure of the π-allyl complex **19** in THF-d_8

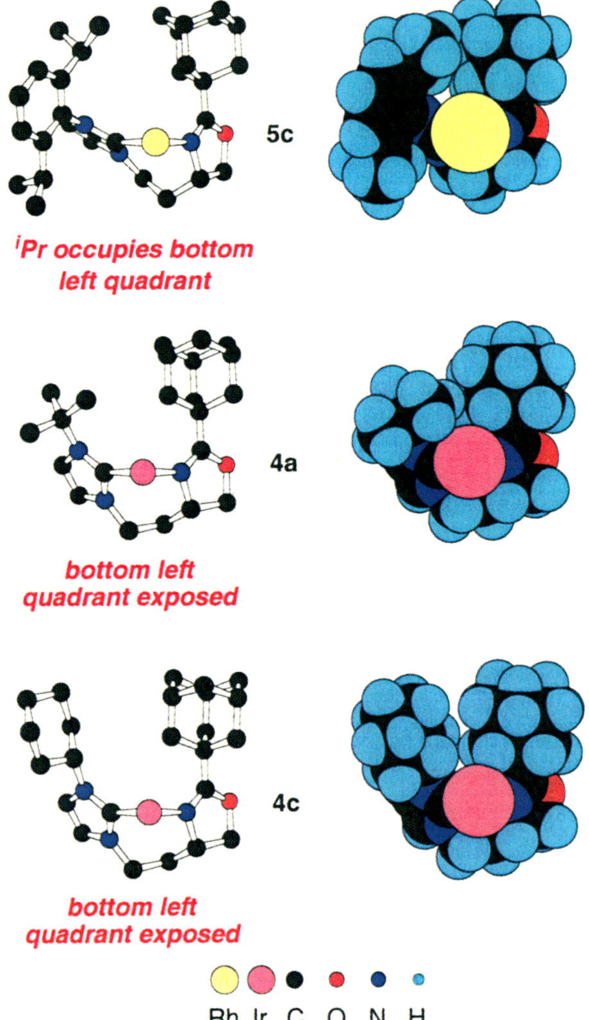
5c
^{i}Pr occupies bottom left quadrant
4a
bottom left quadrant exposed
4c
bottom left quadrant exposed
Rh
Ir
C
O
N
H

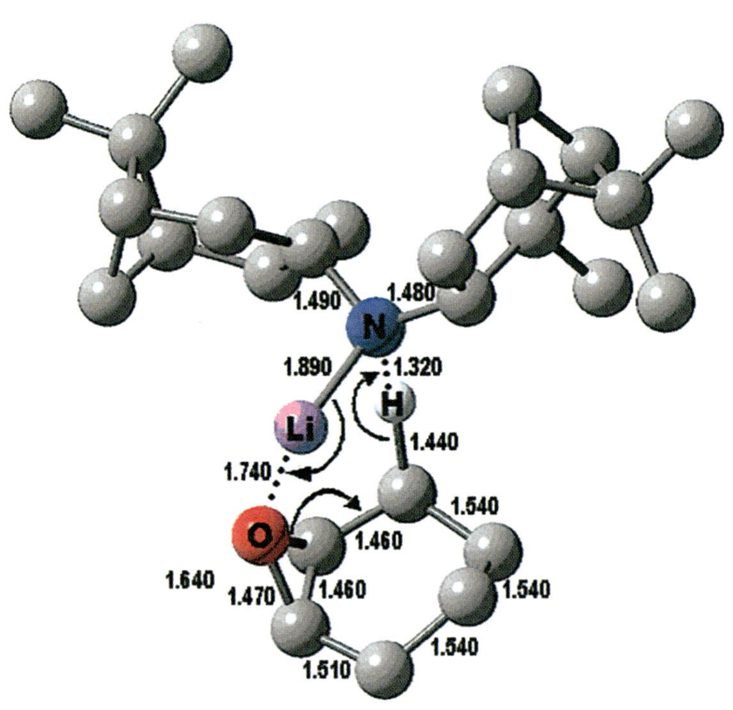

1.490
1.480
N
1.890
1.320
H
Li
1.440
1.740
1.540
O
1.460
1.640
1.470
1.460
1.540
1.540
1.510

Chapter 6

Asymmetric Autocatalysis and the Origin of Homochirality of Biomolecules

Kenso Soai, Itaru Sato, and Takanori Shibata

Department of Applied Chemistry, Faculty of Science, Tokyo University of Science, Kagurazaka, Shinjuku-ku, Tokyo 162–8601, Japan

When a chiral compound operates as a chiral catalyst for its own production, the process is defined as asymmetric autocatalysis. In the enantioselective addition of diisopropylzinc to nitrogen-containing aryl aldehydes, asymmetric autocatalysis has been realized. 3-Pyridyl-, 5-pyrimidyl-, and 3-quinolylalkanols are efficient chiral catalysts for the enantioselective isopropylation of the corresponding aldehydes, and the alkanols themselves are obtained in high yields and in high enantiomeric excess (ee). Asymmetric autocatalyst with extremely low ee value automultiplies with significant amplification of the ee to afford the (same) products with extremely high ee values. In the presence of chiral molecules with only a slight enantiomeric imbalance, which can be induced by chiral physical factors, organic molecules with extremely high ee are obtained using an asymmetric autocatalytic system. Moreover, absolute asymmetric synthesis, i.e., the reaction of pyrimidine carbaldehyde and diisopropylzinc in the absence of chiral molecules, yields enantioenriched (*S*)-or (*R*)-alkanols with a stochastic distribution.

Catalytic asymmetric synthesis is a very efficient protocol for the synthesis of chiral compounds, because a high yield of enantiomerically enriched molecules can be obtained using a small amount of enantiomerically enriched molecules (chiral catalysts). In fact, many types of selective chiral catalyst have been reported for various asymmetric reactions, including reduction, oxidation, and carbon–carbon forming reactions.[1] The structures of the initial asymmetric catalysts and the chiral products in these reactions are usually very different.

In contrast, in asymmetric autocatalysis,[2,3] the structures of an asymmetric catalyst and the chiral product are the same, which means that the chiral product acts as a chiral catalyst for its own production, and that enantioselective automultiplication proceeds without the aid of other chiral compounds (auxiliaries) (see Scheme 1). Mathmatical mechanism of asymmetric autocatalysis was proposed, but no concrete examples of asymmetric autocatalytic reactions have been disclosed.[4] Therefore, the discovery of chiral molecules that have asymmetric autocatalytic abilities is a challenging and fascinating topic in organic chemistry.

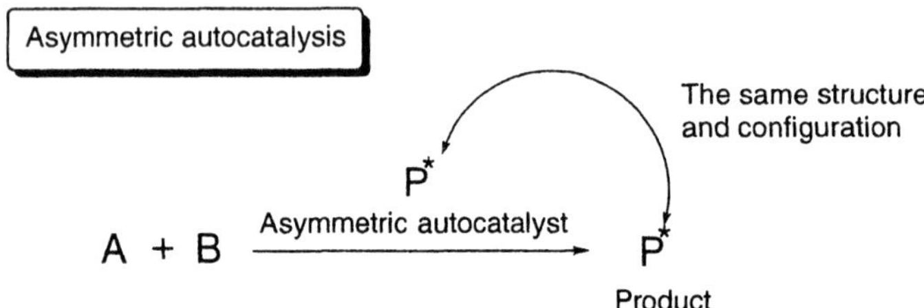

Scheme 1. The principle of asymmetric autocatalysis.

Alberts and Wynberg reported on asymmetric autoinduction.[5a] However, this type of reaction needs to be distinguished from asymmetric autocatalysis. In asymmetric autoinduction,[5] a chiral product forms a chiral complex with a catalyst, and therefore, the process is related to asymmetric induction, but the compound itself cannot operate as a chiral catalyst.

Discovery and Development of Asymmetric Autocatalysis

In 1990, we discovered the first asymmetric autocatalysis reaction in the enantioselective isopropylation of aldehydes.[6] When the addition of diisopropylzinc (i-Pr_2Zn) to pyridine-3-carbaldehyde using a catalytic amount of (*S*)-2-methyl-1-(3-pyridyl)-1-propanol **1** with 86% ee was examined, it was found that (*S*)-pyridyl alkanol **1** was newly formed in a 67% yield, which possessed the same configuration (*S* enantiomer; 35% ee) as the catalyst. Chiral isopropylzinc alkoxide formed *in situ* from alkanol **1** and i-Pr_2Zn operated as a true catalyst.

Other than nitrogen-containing alcohols, ferrocenyl alkanol **3** and diol **4** have also been found to be asymmetric autocatalysts.[7,8]

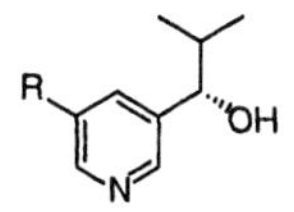

1: R=H
2: R=CON(i-Pr)$_2$

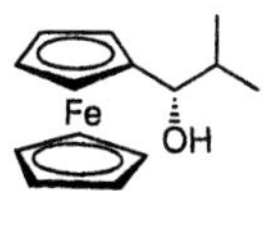

3

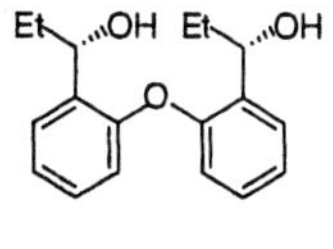

4

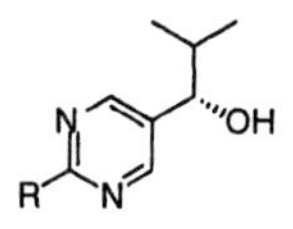

5: R=H
6: R=Me

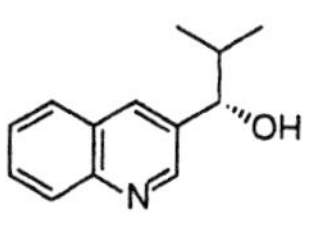

7

Highly Enantioselective Asymmetric Autocatalysis

The above-mentioned alcohols operate as asymmetric autocatalysts. However, their enantioselectivity is not very high. The ee of the alcohols formed is lower than that of the catalysts. Chiral pyrimidyl alkanols were found to exhibit the first highly enantioselective asymmetric autocatalysis.[9] When chiral (*S*)-pyrimidyl alkanol **5** with a 93% ee was used as an asymmetric autocatalyst in the enantioselective addition of i-Pr_2Zn to pyrimidine-5-carbaldehyde, the (*S*)-pyrimidyl alkanol **5** formed had an ee of 90%. When a chiral (*S*)-pyrimidyl alkanol with a 2-methyl substituent **6** having >99.5% ee was used, then the ee of the newly formed (*S*)-alkanol **6** reached a value of 98%.

This asymmetric autocatalysis is superior to conventional (non-autocatalytic) asymmetric synthesis for the following reasons: (1) It is not necessary to have an asymmetric catalyst with a structure different from that of the chiral product. (2) The chiral product serves as an asymmetric catalyst for its own production, and therefore, the chiral product automultiplied itself exponentially. (3) The catalyst does not deteriorate, because the catalyst formed continuously as the reaction proceeded, and (4) separation of the chiral product from the (auto)catalyst is not necessary.

Chiral 3-quinolyl alkanol **7**[10] and chiral 5-carbamoyl-3-pyridyl alkanols[11] have also been shown to be highly enantioselective asymmetric autocatalysts. (*S*)-3-quinolyl alkanol **7** with a 94% ee catalyzes the enantioselective addition of i-Pr_2Zn to quinoline-3-carbaldehyde to yield (*S*)-alkanol 7 with a 94% ee, having the same configuration as the catalyst.[10]

Introduction of a carbamoyl group at the 5-position of the pyridine ring drastically increases the enantioselectivity of 3-pyridyl alkanol acting as an asymmetric autocatalyst.[11] Chiral (*S*)-5-carbamoyl-3-pyridyl alkanol **2** (94% ee) catalyzes the enantioselective addition of i-Pr_2Zn to 5-carbamoylpyridine-3-carbaldehyde to yield (*S*)-alkanol **2** with an ee up to 86%. Substituents on the nitrogen atom of the carbamoyl group play a pivotal role in the asymmetric induction, and a pyridyl alkanol possessing isopropyl groups on the nitrogen atom is the most highly enantioselective asymmetric autocatalyst.

A Near Perfect Asymmetric Autocatalyst: 2-Alkynyl-5-pyrimidyl Alkanol

As a result of studying the effect of substituents on the pyrimidine ring, the 2-position was found to be the most important position for asymmetric induction. Among the various substituents examined, ethynyl[12] and alkenyl[13] groups were found to be the best (see Scheme 2): When (*S*)-1-(2-*tert*-butylethynyl-5-pyrimidyl)-2-methylpropanol **9** with >99.5% ee was used as an asymmetric autocatalyst in the addition of i-Pr_2Zn to 2-(*tert*-butylethynyl)pyrimidine-5-carbaldehyde **8** in cumene at 0 °C, then (*S*)-pyrimidyl alkanol **9** was obtained quantitatively in near perfect enantioselectivity (>99% isolated yield, and >99.5% ee).[12] Next, we examined a consecutive reaction, where the (*S*)-alkanol **9** that was obtained was further used as a catalyst for a new reaction. As a result, even the tenth such reaction proceeded in a quantitative manner in near-perfect enantioselectivity. A sixfold increase in (*S*)-pyrimidyl alkanol **9** was afforded by each reaction, and thus, the amplification factor of ten successive asymmetric autocatalytic reactions reached a value of 6^{10} (*ca.* 60,000,000). In contrast, if the (*R*)-pyrimidyl alkanol **9** with >99.5% ee was used as an asymmetric autocatalyst instead of the (*S*)-**9**, then the resulting compound (*R*)-**9** with >99.5% ee was obtained in >99% isolated yield.

8 + i-Pr_2Zn → (*S*)-**9** [catalyst: (*S*)-**9**, >99.5% ee, 20 mol%; cumene, 0 °C]

consecutive runs	yield, ee
1st run	>99%, >99.5% ee
2nd run	>99%, >99.5% ee
:	: :
10th run	>99%, >99.5% ee

Scheme 2. Near perfect asymmetric autocatalysis, in which the product is used as the asymmetric autocatalyst for the next run.

Asymmetric Autocatalysis with Amplification of the Enantiomeric Excess

When an asymmetric autocatalyst with a low ee is used, then a product with a higher ee value can be obtained. Using the (*S*)-pyrimidyl alkanol **5** with only a 2% ee value as a catalyst in the reaction between *i*-Pr_2Zn and the aldehyde **10**, compound (*S*)-**5**, which includes the catalyst and a newly formed alkanol **5** was obtained with an amplified ee of 10% (see Figure 1).[14] In this case, the asymmetric autocatalytic reactions we examined were reactions where the product obtained from one reaction served as the asymmetric autocatalyst for the next. Successive reactions providing an amplification of the ee value are one of the advantages of asymmetric autocatalysis, and this can never be realized in *non*-autocatalytic asymmetric amplification. The ee of the obtained product **5** increased to 57% ee, 81% ee, and 88% ee, over consecutive reactions.[14] This is an unprecedented and unique asymmetric reaction pathway, which includes asymmetric autocatalysis and asymmetric amplification. As shown in Figure 1, during the first four consecutive asymmetric autocatalytic reactions, the initial catalyst (*S*)-**5** increases 239-fold. On the other hand, (*R*)-**5** increases by only 16-fold.

It should be noted that, even if an asymmetric autocatalyst with an extremely low ee was used as the initial catalyst, an almost enantiomerically pure product (asymmetric autocatalyst) can be obtained by consecutive asymmetric reactions.[15] In practice, only a slightly (*S*)-enriched **9** with a 0.00005% ee value (*S* isomer:*R* isomer = 50.000025:49.999975) was used as the initial catalyst, and an almost enantiomerically pure **9** resulted after only three consecutive enantioselective isopropylations of aldehyde **8** (see Scheme 3). During these asymmetric autocatalysis reactions, the slight excess of (*S*)-**9** in the initial catalyst was automultiplied by a factor of *ca.* 630,000, while the (*R*)-**9** was automultiplied by a factor of less than 1,000.

Chiral isopropylzinc alkoxides formed *in situ* from the (*S*)- and (*R*)-pyrimidyl alkanols **9** and *i*-Pr_2Zn probably also work as asymmetric autocatalysts. A sigmoid curve in the relationship between the yield of the product **9** and the reaction time suggests that the reaction is autocatalytic and that reaction proceeds via second order with isopropylzinc alkoxide of pyrimidyl alkanol **9**.[16] In the amplification of the ee, an inhibition mechanism can suppress the activity of the minor enantiomer of the catalyst. Kagan discussed this aspect in non-autocatalytic reactions as arising from aggregations of chiral ligands.[17] We also postulate that aggregation of the catalyst occurs. However, clarification of the inhibition mechanism in the present asymmetric autocatalysis requires further investigation.

Not only pyrimidyl alkanols, but 3-quinolyl alkanols and 5-carbamoyl-3-pyridyl alkanols are also asymmetric autocatalysts providing an amplification of the ee value.[18]

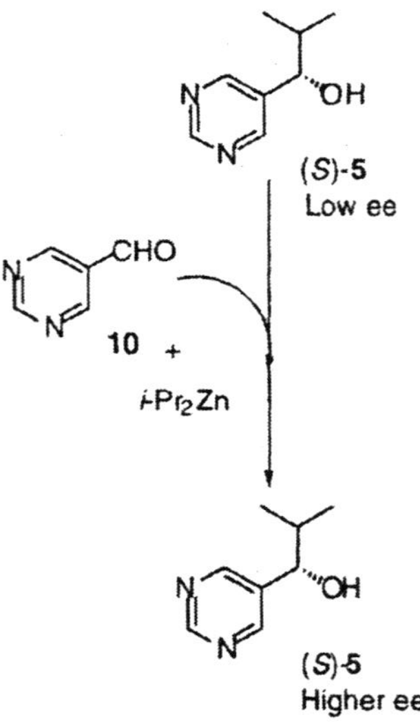

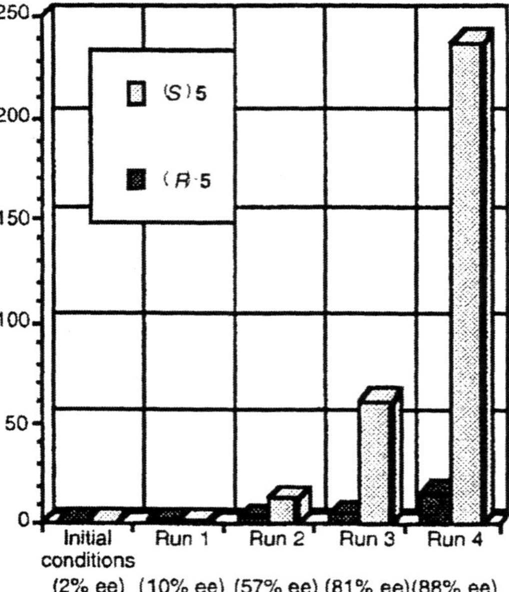

Figure 1. The first asymmetric autocatalysis with amplification of ee

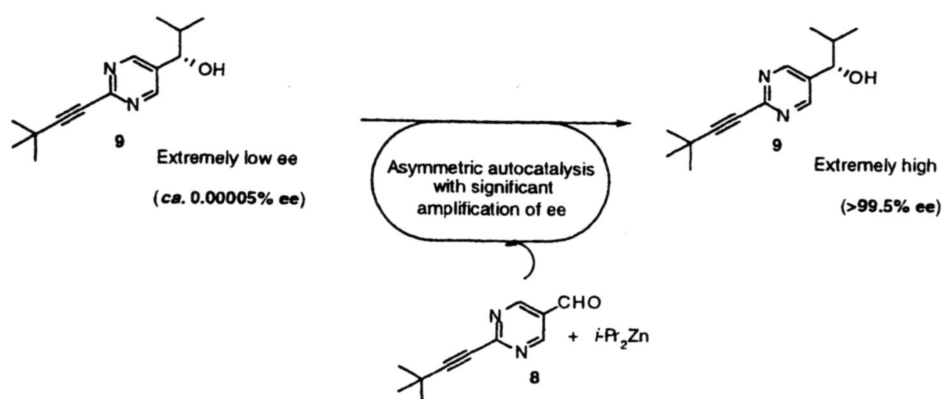

*Scheme 3. Amplification of extremely low ee values in the asymmetric autocatalysis of **9**.*

Asymmetric Autocatalysis Initiated by Various Chiral Organic Molecules

If the alkylation of pyrimidine-5-carbaldehyde with diisopropylzinc is conducted in the presence of a chiral compound with a low ee value (denoted here as a, "chiral initiator"), and the chiral initiator is not a pyrimidyl alkanol, then a slight enantiomeric imbalance is expected to be induced in the initially formed zinc alkoxide of the pyrimidyl alkanol. This small imbalance is drastically amplified by the ensuing asymmetric autocatalysis of the pyrimidyl alkanol, even if it was initially extremely small.[19]

The asymmetric alkylation of pyrimidine-5-carbaldehyde **11** by i-Pr_2Zn was studied when methyl mandelate **12** with a low ee of *ca.* 0.1% was used as the chiral initiator (Scheme 4). In the presence of (*R*)-**12**, (*S*)-pyrimidyl alkanol **6** was obtained with a much higher ee than that of the chiral initiator.[19a] On the other hand, in the presence of (*S*)-**12**, (*R*)-**6** with much higher ee was formed. These results show that small enantiomeric imbalances in the methyl mandelate **12** were recognized and amplified in the resulting pyrimidyl alkanol **6**. Chiral carboxylic acid **13**, amines **14**, and alcohols also work as chiral initiators. It should be noted that this effect is observed even with 2-butanol **15** having *ca.* 0.1% ee, because a slight difference between the methyl and ethyl substituents on the asymmetric carbon of 2-butanol can be recognized by the reaction between pyrimidine-5-carbaldehyde **11** and i-Pr_2Zn.

Can the Mystery of Homochirality be Solved by Asymmetric Autocatalysis?

The enantiomeric ratios of naturally occurring organic molecules are biased towards one enantiomer, such as that seen in L-amino acids or D-ribose. The origin of this homochirality has been a topic of discussion for many years.[20] Not only the chiral substances, but also chiral physical forces have been proposed as being the origin of chiral bias. However, the enantiomeric imbalance induced by these chiral factors is very small, and therefore, without any chiral amplification mechanism, it has not been possible to correlate tiny imbalances in the homochirality of biologically important molecules. On the other hand, this tendency can be explained by enantiomeric amplification of the small fluctuations in a random chance mechanism. It is noteworthy that the asymmetric amplification mechanism is indispensable for making a significant connection between the small fluctuations and a large enantiomeric imbalance. In the latter case, if we examine the random chance mechanism with a suitable enantiomeric amplification method, then the stochastic formation of two enantiomers should be observed.

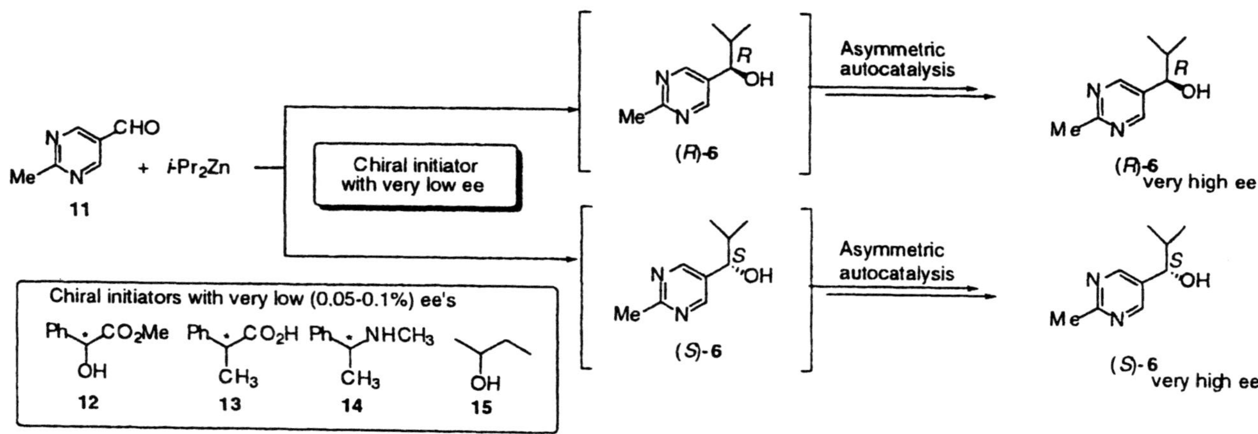

Scheme 4. Asymmetric Autocatalysis Initiated by Various Chiral Molecules

Circularly Polarized Light as an Origin of Chirality

Several physical factors have been proposed as being the origin of chiral organic compounds, and circularly polarized light (CPL) is a potent candidate among these. However, the degree of enantiomeric imbalance induced by CPL[21] is too small to be correlated with the large enantiomeric imbalance observed in molecules found in nature. For example, the asymmetric photolysis of racemic leucine **16** with right-CPL affords L-leucine **16**, but with only a 2% ee (L:D = 51:49).[21a] Similarly, in asymmetric photosynthesis using CPL, chiral [6]helicene **17** can be obtained, but with a very low ee (<0.2%).[21b, c] We have developed a chemical process where these chiral organic compounds, induced by CPL, can be correlated with organic compounds having a very high ee.

In the presence of L-leucine having only 2% ee, the asymmetric alkylation of 2-methylpyrimidine-5-carbaldehyde **11** with i-Pr_2Zn affords (*R*)-pyrimidyl alkanol **6** with increased ee of 21% (Scheme 5).[19a] On the other hand, when D-leucine **16** with a 2% ee was used as the chiral initiator, the (*S*)-pyrimidyl alkanol **6** with an increased ee (26%) was obtained. These imbalances (21 % ee and 26 % ee) can be amplified by the following asymmetric autocatalysis reaction.[19a]

Then, we examined (*P*)-[6]helicene **17** as a chiral initiator. In the presence of (*P*)-[6]helicene **17** with only a 0.13% ee, the reaction between pyrimidine-5-carbaldehyde **8** and i-Pr_2Zn affords (*S*)-pyrimidyl alkanol **9** with 56% ee (Scheme 6).[22] In the presence of (*M*)-[6]helicene **17** with a 0.54% ee, an (*R*)-pyrimidyl alkanol **9** with 62% ee was obtained. In addition, [5]helicene also acts as a chiral initiator.[22] When these results are observed from another point of view, then they can be seen as being the first examples of highly enantioselective synthesis induced by a helically chiral hydrocarbon without any heteroatoms.

Chiral Inorganic Crystals as Origins of Chirality

Morphologically, quartz (SiO_2) is a chiral crystal and both dextrorotatory (*d*) and levorotatory (*l*) enantiomorphs exist in nature. No obvious asymmetric induction by quartz has been reported,[23] although, an asymmetric adsorption of a chiral compound by quartz has been reported. However, the degree of chiral differentiation is very small.[24]

We considered that even quartz, an inorganic chiral crystal, could work as a chiral initiator in our chiral recognition system based on asymmetric autocatalysis. When enantiomorphs induce a slight enantiomeric imbalance in the initially formed zinc alkoxide of pyrimidyl alkanol **9**, which is provided by the reaction between pyrimidine-5-carbaldehyde **8** and i-Pr_2Zn, then the imbalance is amplified by the subsequent asymmetric autocatalysis to yield pyrimidyl alkanol **9** with a very high ee.

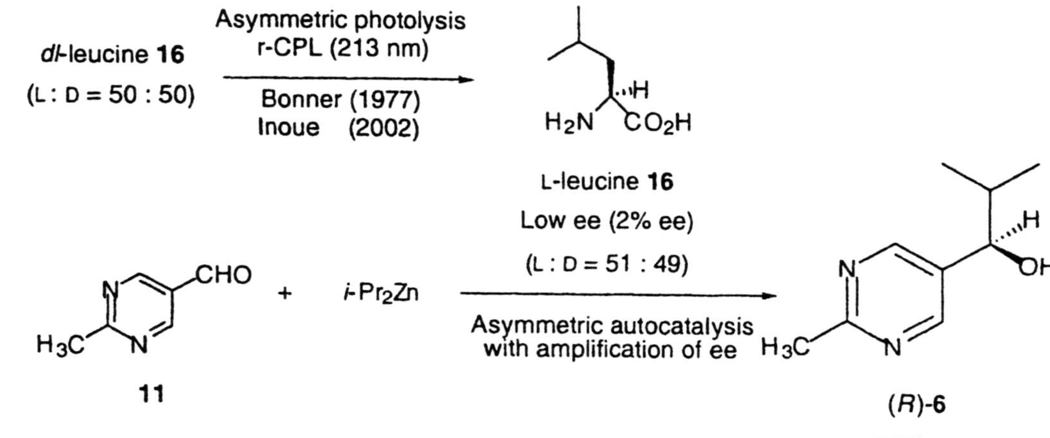

Scheme 5. Asymmetric autocatalysis in the presence of leucine with low ee.

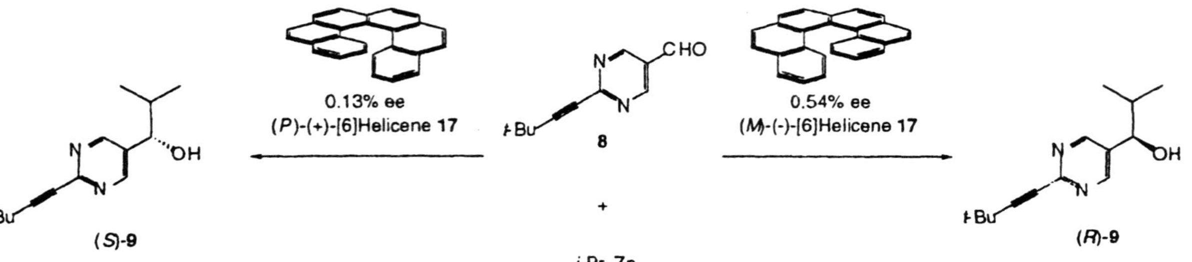

Scheme 6. Asymmetric autocatalysis in the presence of [6]helicene.

We examined the reaction of 2-alkynylpyrimidine-5-carbaldehyde **8** with *i*-Pr_2Zn in the presence of powdered *d*-quartz. As a result, (*S*)-pyrimidyl alkanol **9** with a very high ee (97%) was obtained in a 95% yield (Scheme 7).[25] Conversely, in the presence of *l*-quartz, (*R*)-**9** with a 97% ee was formed in a 97% yield. These results clearly show that the enantiomorphs of quartz can determine the absolute configuration of the resulting pyrimidyl alkanol. Thus, this is the first example of a chiral inorganic crystal inducing chirality in an organic compound with a high ee.

Next, we studied sodium chlorate ($NaClO_3$) as a chiral initiator, because the stirred crystallization of racemic sodium chlorate affords either *d*- or *l*-enantiomerically-enriched ionic crystals.[26] We found that *d*-$NaClO_3$ induced the (*S*)-pyrimidyl alkanol **9** with a 98% ee in the enantioselective isopropylation of aldehyde **8**.[27] On the other hand, *l*-$NaClO_3$ afforded (*R*)-**9** with a 98% ee.

Stochastic Production of *S*- and *R*-Enantiomers without a Chiral Source: Absolute Asymmetric Synthesis in Combination with Asymmetric Autocatalysis

When a nucleophilic addition to an achiral aldehyde in the absence of any chiral environment obeys the rules of statistics, i.e., the probability of *re*-face and *si*-face attack is 1:1, then one would obtain the product as a so-called, "racemate". However, based on statistical theory, in almost all cases the racemate does not contain exactly the same number of *R*- and *S*-enantiomers.

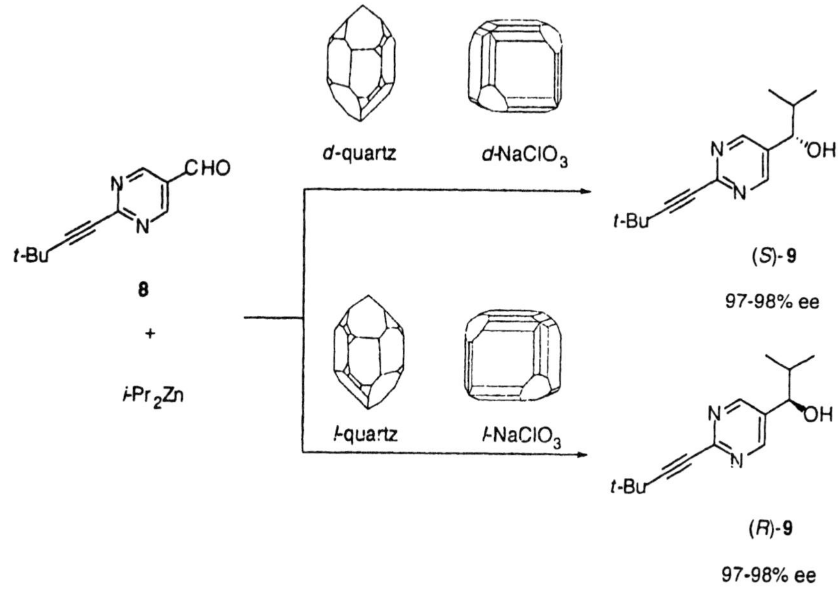

Scheme 7. Asymmetric autocatalysis in the presence of chiral inorganic crystals.

This is attributed to random fluctuations. It should be noted that, in a large number of molecules, the ee is extremely small and cannot be detected even by the modern ee analysis techniques. Mill calculated the expected small fluctuations in the ratios of two enantiomers, and concluded that the fluctuations become more and more obvious as the number of the molecules becomes smaller. [If a group of 100,000 chiral molecules are produced under conditions where the probabilities for the formation of two enantiomers is equal, then statistically, half the groups will contain an excess of more than 212 molecules of one enantiomer (0.21% ee)][28] If the initial small fluctuations in chirality are amplified by the asymmetric autocatalysis mechanism, then one would obtain an enantioenriched compound with a detectable ee. Mislow published a comprehensive review and commentary on absolute asymmetric synthesis.[29]

Indeed, we obtained enantioenriched pyrimidyl alkanols with either *S* or *R* configurations from the reaction of pyrimidinecarbaldehyde and diisopropylzinc without the addition of a chiral substance.[30]

The reaction of 2-alkynylpyrimidine-5-carbaldehyde **8** with i-Pr_2Zn in a mixed solvent of ether and toluene affords the optically active (*R*)-pyrimidyl alkanol **9** with 75% ee.[31] The ee value of the obtained pyrimidyl alkanol **9** was high enough to be determined by HPLC using a chiral stationary phase. To examine the statistical distribution of the absolute configuration of the formed **9**, an additional 36 experiments were conducted using new, clean equipment (e.g., glassware, syringe, needle, and stirrers.) under the same reaction conditions. In all cases, the optically active pyrimidyl alkanol **9** with either *S* or *R* configurations was obtained. As shown in Figure 2, the absolute configurations of the pyrimidyl alkanol **9** formed show an approximate stochastic distribution (formation of *S* = 19 times, and formation of *R* = 18 times).[31] A slight enantiomeric imbalance in the initially formed pyrimidyl alkanol was considered to have been amplified by the in situ consecutive asymmetric autocatalysis, affording pyrimidyl alkanol with the ee level high enough to be detectable. The stochastic distribution of the absolute configurations of the product, observed for the first time, strongly suggests that the reaction was an absolute (spontaneous) asymmetric synthesis.

Conclusion

Chiral zinc alkoxides of 5-pyrimidyl, 3-pyridyl, and 3-quinolyl alkanols operate as asymmetric autocatalysts, which automultiply in the enantioselective addition of *i*-Pr_2Zn to the corresponding aldehydes. In particular, 2-alkynyl-5-pyrimidyl alkanol realizes practically perfect enantioselectivity (>99.5% ee) in a quantitative reaction (>99% yield). In principle, consecutive asymmetric autocatalysis can infinitely automultiply chiral compounds with a high ee. Asymmetric autocatalysis is a superior asymmetric synthetic method to conventional (non-autocatalytic) methods, as no chiral catalyst is needed other than the compound itself, the product yield increases exponentially, and separation of the product from the catalyst is unnecessary.

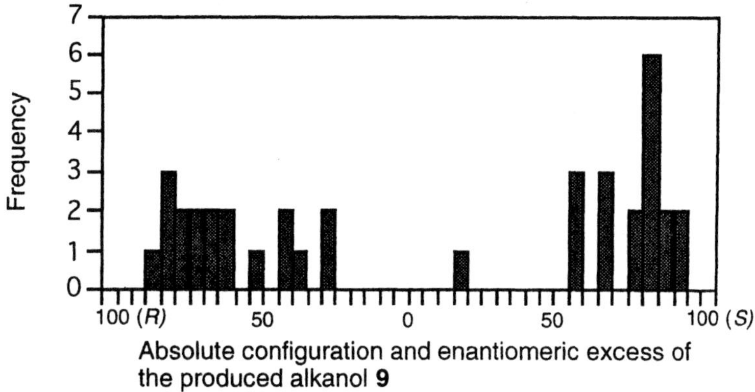

*Figure 2. Stochastic production of chiral **9** in the reaction of achiral aldehyde **8** and* i-Pr_2Zn.

Starting from asymmetric autocatalysts with an extremely low ee, consecutive asymmetric autocatalysis enables the automultiplication of the product (catalyst) with significant amplification of ee without the need for any other chiral auxiliaries. Moreover, we can initiate asymmetric autocatalysis using chiral compounds (chiral initiator) other than pyrimidyl alkanol, even those having a very low ee. The absolute configuration of a slightly enantiomerically enriched isomer in a chiral initiator determines the chirality of the pyrimidyl alkanol. It is noteworthy that CPL-induced chiral compounds, such as leucine and [6]helicene, can also work as chiral initiators. Thus, asymmetric autocatalysis with an amplification of ee serves as a mediator that correlates with a slight enantiomeric imbalance in the molecules induced by CPL with a high enantiomeric imbalance in the resulting organic molecules. Furthermore, inorganic crystals, such as quartz and sodium chlorate, are also chiral initiators, and the absolute configuration of the obtained pyrimidyl alkanol is dependent on the excess of the dextrorotatory (*d*) or levorotatory (*l*) enantiomorph.

Asymmetric induction has been observed even in the absence of a chiral initiator. The probability of absolute configurations of the pyrimidyl alkanol **2** show an approximate stochastic distribution.

We have discovered and characterized highly enantioselective asymmetric autocatalytic reactions and asymmetric autocatalysis with amplification of ee. We believe that our system will elucidate the mystery of homochirality in naturally occurring compounds.[32]

Acknowledgment

Special gratitude is expressed to all of our co-workers, whose names appear in the reference list. Financial support from the Ministry of Education, Culture,

Sports, Science and Technology (MEXT), the New Energy and Industrial Technology Development Organization (NEDO), and the Daicel Award in Synthetic Organic Chemistry is gratefully acknowledged.

Literature Cited

1. (a) *Comprehensive Asymmetric Catalysis*; Jacobsen, E.; Pfaltz, A.; Yamamoto, H., Ed.; Springer-Verlag: Heidelberg, 1999. (b) *Catalytic Asymmetric Synthesis, Second Edition*; Ojima, I., Ed.; John Wiley, New York, 2000.
2. (a) Soai, K.; Shibata, T. In *Advances in Biochirality*; Pályi, G.; Zucchi, C.; Caglioti, L., Ed.; Elsevier: Amsterdam, 1999; Chapter 11. (b) Soai, K.; Shibata, T.; Sato, I. *Acc. Chem. Res.* **2000**, *33*, 382. (c) Soai, K. *Enantiomer*, **1999**, *4*, 591. (d) Soai, K.; Shibata, T. *Yuki Gosei Kagaku Kyokaishi* (*J. Synth. Org. Chem. Jpn.*), **1997**, *55*, 994. (e) Soai, K.; Shibata, T. In *Catalytic Asymmetric Synthesis, Second Edition*; Ojima, I., Ed.; Chapter 9, Wiley, New York, 2000. (f) Soai, K.; Shibata, T.; Sato, I. *Nippon Kagaku Kaishi* (J. Chem. Soc. Jpn., Chem. Ind. Chem.), **2001**, 141. (g) Soai, K.; Sato, I.; Shibata, T. *Chem. Record* **2001**, *1*, 321. (h) Soai, K.; Sato, I. *Chirality* **2002**, *14*, 548. (i) Soai, K. In *Fundamentals of Life*, Pályi, G.; Zucchi, C.; Caglioti, L., Ed.; pp. 427-, Elsevier: Paris, 2002. (j) Soai, K.; Sato, I.; Shibata, T. *Yuki Gosei Kagaku Kyokaishi* (*J. Synth. Org. Chem. Jpn.*), **2002**, *60*, 668. (k) Soai, K.; Sato, I. *Viva Origino*, **2002**, *30*, 186.
3. (a) Bolm, C.; Bienewald, F.; Seger, A. *Angew. Chem., Int. Ed. Engl.* **1996**, *35*, 1657. (b) Avalos, M.; Babiano, R.; Cintas, P.; Jiménez, J. L.; Palacios, J. C. *Chem. Commun.* **2000**, 887. (c) Buschmann, H.; Thede, R.; Heller, D. *Angew. Chem. Int. Ed.* **2000**, *39*, 4033. (d) Todd, M. H. *Chem. Soc. Rev.* **2002**, *31*, 211.
4. (a) Frank, F. C. *Biochim. Biophys. Acta* **1953**, *11*, 459. Reviews: (b) Calvin, M. *Chemical Evolution*; Oxford University, London, 1969; Chapter 7. (c) Wynberg, H. *J. Macromol. Sci.-Chem. A* **1989**, *26*, 1033.
5. (a) Alberts, A. H.; Wynberg, H. *J. Am. Chem. Soc.* **1989**, *111*, 7265. (b) Soai, K.; Inoue, Y.; Takahashi, T.; Shibata, T. *Tetrahedron* **1996**, *52*, 13355. (c) Shibata, T.; Takahashi, T.; Konishi, T.; Soai, K. *Angew. Chem. Int. Ed. Engl.* **1997**, *36*, 2458. (d) Danda, H.; Nishikawa, H.; Otaka, K. *J. Org. Chem.* **1991**, *56*, 6740.
6. Soai, K.; Niwa, S.; Hori, H. *J. Chem. Soc. Chem. Commun.* **1990**, 982.
7. Soai, K.; Hayase, T.; Takai, K. *Tetrahedron: Asymmetry* **1995**, *6*, 637.
8. Soai, K.; Hayase, T.; Shimada, C.; Isobe, K. *Tetrahedron: Asymmetry* **1994**, *5*, 789.
9. Shibata, T.; Morioka, H.; Hayase, T.; Choji, K.; Soai, K. *J. Am. Chem. Soc.* **1996**, *118*, 471.
10. Shibata, T.; Choji, K.; Morioka, H.; Hayase, T.; Soai, K. *Chem. Commun.* **1996**, 751.
11. Shibata, T.; Morioka, H.; Tanji, S.; Hayase, T.; Kodaka, Y.; Soai, K. *Tetrahedron Lett.* **1996**, *37*, 8783.

12. Shibata, T.; Yonekubo, S.; Soai, K. *Angew. Chem. Int. Ed.* **1999**, *38*, 659.
13. Sato, I.; Yanagi, T.; Soai, K. *Chirality* **2002**, *14*, 166.
14. Soai, K.; Shibata, T.; Morioka, H.; Choji, K. *Nature (London)* **1995**, *378*, 767.
15. Sato, I.; Urabe, H.; Ishiguro, S.; Shibata T.; Soai. K. *Angew. Chem. Int. Ed.* **2003**, *42*, 315.
16. (a) Sato, I.; Omiya, D.; Tsukiyama, K.; Ogi, Y.; Soai, K. *Tetrahedron: Asymmetry* **2001**, *12*, 1965. (b) Sato, I.; Omiya, D.; Igarashi, H.; Kato, K.; Ogi, Y.; Tsukiyama, K.; Soai, K. *Tetrahedron: Asymmetry* **2003**, *14,* 975. (c) Blackmond, D. G.; McMillan, C. R.; Ramdeehul, S.; Schorm, A.; Brown, J. M. *J. Am. Chem. Soc.* **2001**, *123*, 10103.
17. Reviews: (a) Girard, C.; Kagan, H. B. *Angew. Chem. Int. Ed.* **1998**, *37*, 2922. (b) Avalos, M.; Babiano, R.; Cintas, P.; Jiménez, J. L.; Palacios, J. C. *Tetrahedron: Asymmetry* **1997**, *8*, 2997. (c) Fenwick, D. R.; Kagan, H. B. In *Topics in Stereochemistry*; Denmark, S. E., Ed.; Wiley: New York, 1999; Vol. 22, pp 257. (d) Bolm, C. *Advanced Asymmetric Synthesis*; Stephenson, G. R., Ed.; Blackie: Glasgow, 1996, pp 9. (e) Green, M. M.; Park, J.-W.; Sato, T.; Lifson, S,; Selinger, R. L. B.; Seliger, J. V. *Angew. Chem. Int. Ed.* **1999**, *38*, 3138.
18. (a) Shibata, T.; Choji, K.; Hayase, T.; Aizu, Y.; Soai, K. *Chem. Commun.* **1996**, 1235. (b) Tanji, S.; Kodaka, Y.; Ohno, A.; Shibata, T.; Sato, I.; Soai, K. *Tetrahedron: Asymmetry* **2000**, *11*, 4249.
19. (a) Shibata, T.; Yamamoto, J.; Matsumoto, N.; Yonekubo, S.; Osanai, S.; Soai, K. *J. Am. Chem. Soc.* **1998**, *120*, 12157. (b) Sato, I.; Omiya, D.; Saito, T.; Soai, K. *J. Am. Chem. Soc.* **2000**, *122*, 11739. (c) Tanji, S.; Ohno, A.; Sato, I.; Soai, K. *Org. Lett.* **2001**, *3*, 287. (d) Sato, I.; Kadowaki, K.; Ohgo, Y.; Soai, K. Ogino, H. *Chem. Commun.* **2001**, 1022.(e) Sato, I.; Osanai, S.; Kadowaki, K.; Sugiyama, T.; Shibata, T.; Soai, K. *Chem. Lett.* **2002**, 168.(f) Sato, I.; Kadowaki, K.; Urabe, H.; Jung, J.-H.; Ono, Y.; Shinkai. S.; Soai, K. *Tetrahedron Lett.* **2003**, *44*, 721. (g) Sato, I.; Ohno, A.; Aoyama, Y.; Kasahara, T.; Soai, K.; *Org. Biomol. Chem.* **2003**, *1*, 244. (h) Sato, I.; Matsueda, Y.; Kadowaki, K.; Yonekubo, S.; Shibata, T.; Soai' K. *Helv. Chim. Acta.* **2002**, *85*, 3383.
20. (a) Mislow, K.; Bickart, P. *Isr. J. Chem.* **1976/1977**, *15*, 1. (b) Siegel, J. S. *Chirality* **1988**, *10*, 24. (c) Kondepudi, D. K., Nelson, G. W. *Nature* **1985**, *314*, 438. (d) Weissbuch, I.; Addadi, L.; Leiserowitz, L.; Lahav, M. J. Am. Chem. Soc. 1998, 110, 561. Reviews: (e) Mislow, K. *Collect. Czech. Chem. Commun.* **2003**, *68*, 849. (f) Eschenmoser, A. *Science* **1999**, *284*, 2118. (g) Keszthelyi, L. *Q. Rev. Biophys.* **1995**, *28*, 473. (h) Bonner, W. A. *Orig. Life Evol. Biosphere* **1991**, *21*, 59. (i) Mason, S. F.; Tranter, G. E. *Proc. R. Soc. Lond. A* **1985**, *397*, 45. (j) Avetisov, V.; Goldanskii, V. I. *Proc. Natl. Acad. Sci. USA* **1996**, *93*, 11435. (k) Avalos, M.; Babiano, R.; Cintas, P.; Jiménez, J. L.; Palacios, J. C. *Tetrahedron: Asymmetry* **2000**, *11*, 2845. (l) Feringa, B. L.; van Delden, R. A. *Angew. Chem., Int. Ed. Engl.* **1999**, *38*, 3418. (m) Mason, S. *Chem. Soc. Rev.* **1988**, *17*, 347.
21. (a) Flores, J. J.; Bonner, W. A.; Massey, G. A. *J. Am. Chem. Soc.* **1977**, *99*, 3622. (b) Moradpour, A.; Nicoud, J. F.; Balavoine, G.; Kagan, H.;

Tsoucaris. G. *J. Am. Chem. Soc.* **1971**, *93*, 2353. (c) Bernstein, W. J.; Calvin, M.; Buchardt, O. *J. Am. Chem. Soc.* **1972**, *94*, 494. (d) Inoue, Y.; Tsuneishi, H.; Hakushi, T.; Yagi, K.; Awazu, K.; Onuki, H. *Chem. Commun.* **1996**, 2627. Review: (e) Buchardt, O. *Angew. Chem. Int. Ed. Engl.* **1974**, *13*, 179. (f) Inoue, Y.; *Chem. Rev.* **1992**, *92*, 741.

22. Sato, I.; Yamashima, R.; Kadowaki, K.; Yamamoto, J.; Shibata, T.; Soai, K. *Angew. Chem. Int. Ed.* **2001**, *40*, 1096.
23. Schwab, G. M.; Rudolph, L. *Naturwissenschaften* **1932**, *20*, 363. Terent ev, A. P.; Klabunovskii, E. I. *Sbornik Statei Obshchei Khim.* **1953**, *2*, 1598; *Che. Abstr.* **1955**, *49*, 5262g. The reproducibility of these reports of the very low asymmetric induction using quartz was disproved in later careful experiments by Amariglio *et al.* Amariglio, A.; Amariglio, H.; Duval, X. *Helv. Chim. Acta* **1968**, *51*, 2110.
24. Bonner, W. A.; Kavasmaneck, P. R.; Martin, F. S.; Flores, J. J. *Science* **1974**, *186*, 143.
25. (a) Soai, K.; Osanai, S.; Kadowaki, K,; Yonekubo, S.; Shibata, T.; Sato, I. *J. Am. Chem. Soc.* **1999**, *121*, 11235. (b) For the use of helical silica as a chiral trigger. Sato, I.; Kadowaki, K.; Urabe, H.; Hwa Jung, J.; Ono, Y.; Shinkai, S.; Soai, K. *Tetrahedron Lett.* **2003**, *44*, 721.
26. Kondepudi, D. K.; Kaufman, R. J.; Singh, N. *Science* **1990**, *250*, 975.
27. Sato, I.; Kadowaki, K.; Soai, K. *Angew. Chem. Int. Ed.* **2000**, *39*, 1510.
28. Mills, W. H. *Chem. & Ind.* **1932**, 750.
29. (a) For an excellent comprehensive review and commentary, see ref. 20e. (b) Commentary: Siegel, J. S. *Nature* **2002**, *419*, 346-347.
30. (a) Soai, K.; Shibata, T.; Kowata, Y. *Japan Kokai Tokkyo Koho* 9-268179 (1997). (Application date: February 1, 1996 and April 18, 1996). An anstract of JP9268179 is available from the Europian Patent Office (http://ep.espacenet.com). (b) Singleton, D. A.; Vo, L. K. *J. Am. Chem. Soc.* **2002**, *124*, 10010.
31. Soai, K; Sato, I.; Shibata, T.; Komiya, S.; Hayashi, M.; Matsueda, Y.; Imamura, H.; Hayase, T.; Morioka, H.; Tabira, H.; Yamamoto, J.; Kowata, Y. *Tetrahedron: Asymmetry* **2003**, *14*, 185.

Chapter 7

Enantioselective Alkynylation of Aldehydes Catalyzed by New Chiral Catalysts

Albert S. C. Chan, Gui Lu, and Xingshu Li

Open Laboratory of Chirotechnology of the Institute of Molecular Technology for Drug Discovery and Synthesis and Department of Applied Biology and Chemical Technology, The Hong Kong Polytechnic University, Hong Kong, China

A series of amino alcohol ligands of binaphthyl derivatives were synthesized and were found to be effective in the enantioselective alkynylation of aldehydes. A variety of aromatic aldehydes were converted to the corresponding chiral propargylic alcohols in up to 93% e.e. Chiral catalyst systems by virtue of the cooperation of several chiral ligands were also developed. The alkynylation reaction proceeded smoothly to produce the desired products in up to 99.7% e.e.

Chiral propargylic alcohols are versatile building blocks in the synthesis of a broad variety of biologically active and structurally interesting compounds (*1-3*). Two general approaches to prepare optically active propargylic alcohols have been reported starting from either ynones or aldehydes.

ynone + [H] → propargylic alcohol

aldehyde + $H{-}{\equiv}{-}R_2$ → propargylic alcohol

Although direct reductions of α,β–ynones via chirally modified metal hydrides (*4-8*), reductive cleavage of chiral acetylenic acetals (*9*), enzymatic transformations (*10-15*), hydroboration with chiral oxazaborolidines (*16-18*), and transfer hydrogenation (*19*) are catalytic and quite effective, the method of alkynylation of aldehyde has a strategic synthetic advantage because the latter forms a new C-C bond with concomitant creation of a stereogenic center in a single transformation, while in the former the C-C bond and the stereogenic center are formed separately.

Unlike the catalytic enantioselective addition of dialkyl- and alkenylzinc compounds to aldehydes where considerable progress has been made (*20-23*), the current methods for enantioselective alkynylation reactions are still less developed due to either the use of stoichiometric amounts of catalysts, restricted substrates or the formation of considerable amounts of alkylated products (*24-45*).

Yamamoto *et al.*(*26*) found that the reaction of a steroidal aldehyde **1** with stannylacetylene **2c** in the presence of $TiCl_4$ produced the Cram isomer **3** with high diastereoselectivity (at least 85:15). High selectivity also occured in the reaction of **1** with allylstannane and allylsilane.

Krause and Seebach (*27*) prepared R-≡-Ti(OiPr)$_3$ and used it in the alkynylation of aldehydes with only low to moderate diastereoselectivity.

Soai *et al.*(*28,29*) reported the syntheses of alkynyl alcohols by the enantioselective addition of alkynylzinc reagents to aldehydes in the presence of catalytic amounts of amino alcohols and found that (1*S*,2*R*)-(-)-dibutylnorephedrine (**5**, DBNE) gave moderate e.e., while (*S*)-(+)-(1-methylpyrrolidin-2-yl)diphenylmethanol (**6**, MPDPM) afforded 78% e.e.

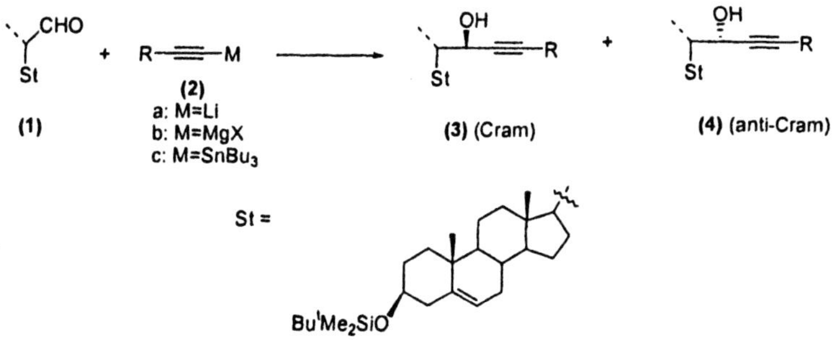

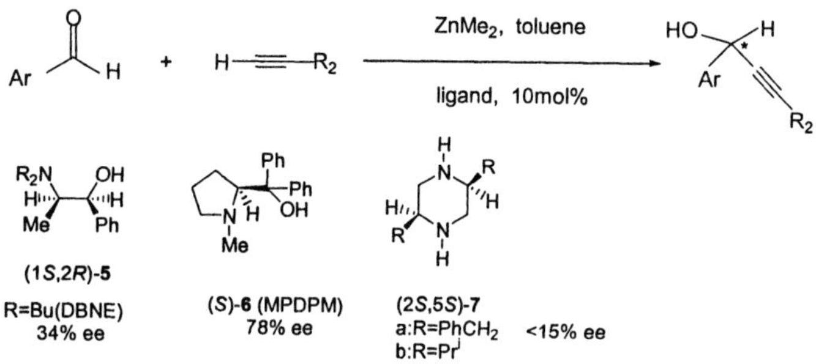

(1) + (2) a: M=Li b: M=MgX c: M=$SnBu_3$ → (3) (Cram) + (4) (anti-Cram)

St = [steroid with Bu^tMe_2SiO]

Ar–CHO + H–≡–R_2 → ($ZnMe_2$, toluene; ligand, 10mol%)

(1*S*,2*R*)-**5** R=Bu(DBNE) 34% ee

(*S*)-**6** (MPDPM) 78% ee

(2*S*,5*S*)-**7** a:R=$PhCH_2$ b:R=Pr^i <15% ee

Tombo and co-workers (*30*) studied the reaction of alkyl and aryl aldehydes with stoichiometric 2-phenylethynylzinc bromide in the presence of (1*R*,2*S*)-(-)-*N*-methylephedrine lithium salt and got the corresponding (*S*)-alcohol in 80% yield with 88% e.e.

Falorni *et al.*(*31,32*) found that ligand **8** showed poor enantioselectivity (16% e.e. and 87% yield) in catalyzing the addition of $(^nBuC{\equiv}C)_2Zn$ to benzaldehyde.

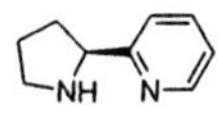

8

Yamaguchi *et al.*(*33*) reported a mild reagent of terminal alkyne, $SnCl_4$ and Bu_3N which alkynylated aldehydes in high yields. Alkynyltrichlorotins were shown to be the reactive species for these reactions.

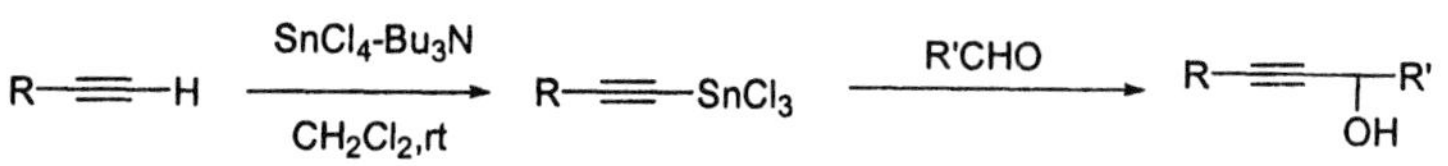

R–≡–H → ($SnCl_4$-Bu_3N, CH_2Cl_2,rt) R–≡–$SnCl_3$ → (R'CHO) R–≡–CH(OH)–R'

Baldoli *et al.*(*34*) added lithium acetylides and ethynyl magnesium bromide to chiral ortho substituted benzaldehyde tricarbonylchromium complexes and obtained alkynyl alcohols in good yields with essentially complete stereoselection (de≥98%).

Corey and co-workers (*35*) described the use of chiral oxazaborolidines as catalysts for the enantioselective addition of alkynylboranes to aldehydes with up to 97% e.e. at low temperature.

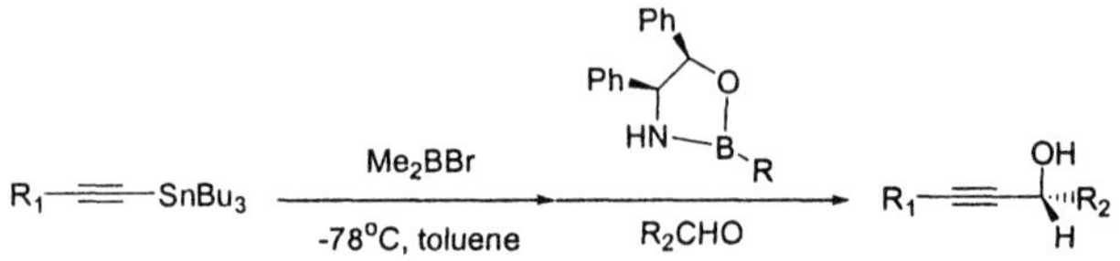

R_1—≡—$SnBu_3$ → (Me_2BBr, -78°C, toluene; oxazaborolidine, R_2CHO) → R_1—≡—CH(OH)R_2

Hoshino *et al.*(*36*) found that tridentate ligands **9a-c** were effective for the direct alkynylation of aldehydes. In the presence of 10 mol% of ligands **9b**, up to 90% e.e. and 95% e.e. were achieved in the addition of bis(2-phenylethynyl)zinc to benzaldehyde and trimethylacetaldehyde respectively. However, the use of ethyl(1-octynyl)zinc resulted a mixture of octynylated and ethylated alcohols.

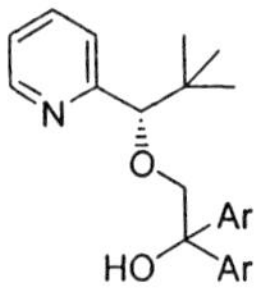

9a: Ar=α-Naphthyl
9b: Ar=β-Naphthyl
9c: Ar=Ph

Recently, Li *et al.*(*37*) studied the use of amino alcohol **10, 11** and their derivatives in the reaction of terminal alkynes with aromatic aldehydes. Up to 85% e.e. and 65-94% yield were obtained in the presence of 10 mol% of chiral ligands.

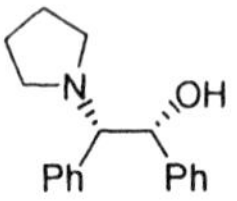

(1*R*,2*S*)-**10** 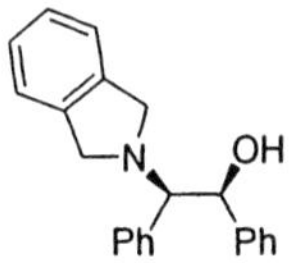 (1*S*,2*R*)-**11**

Carreira's group (*38-42*) reported a system using stoichiometric amount of N-methylephedrine and $Zn(OTf)_2$ to catalyze the reaction of terminal alkynes with a variety of aromatic or aliphatic aldehydes containing α-substituents, very high enantioselectivities were achieved (92-99% e.e.). Recently these investigators also reported the same process in up to 99% e.e. for aliphatic aldehydes using catalytic amounts of $Zn(OTf)_2$ and ligand. The catalytic system was substantially less effective for aromatic substrates.

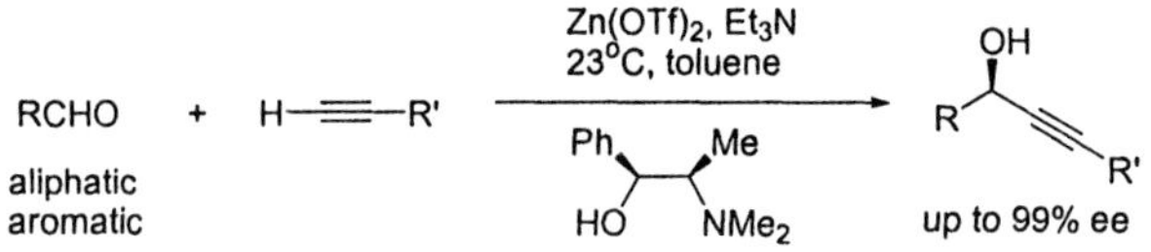

During the preparation of this paper, Pu *et al.*(*44*) reported the use of a (*S*)-BINOL-Ti(OiPr)$_4$ system prepared from a different procedure in the enantioselective reaction of terminal alkynes with aromatic aldehydes (92-98% e.e.). More recently Jiang *et al.*(*45*) prepared a new chiral amino alcohol ligand, (1*S*, 2*S*)-2-*N*,*N*-dimethylamino-1-(*p*-nitrophenyl)-3-(*t*-butyl-dimethylsilyloxy)-propane-1-ol and used it stoichiometrically in the asymmetric alkynylation to give high yields and up to 99% e.e.

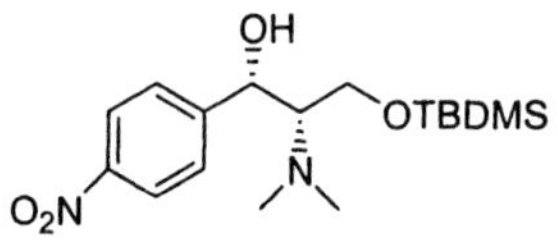

In this paper, we report the results of our recent study (*46-48*) on the enantioselective alkynylation of aldehydes.

Results and Discussions

Our initial envision was that the introduce of a rigid binaphthyl group into amino alcohol might have some structural advantages over traditional amino alcohol in asymmetric catalysis. Following this idea, a series of amino alcohol ligands of binaphthyl derivatives **12-15** were synthesized and they did show good enantioselectivities in the alkylation of benzaldehyde. Further study revealed that one of these ligands, (1*R*,2*S*,3*R*)-**12**, was highly effective in the asymmetric alkynylation of aldehydes. In the presence of dimethylzinc, various aromatic aldehydes were converted to the corresponding chiral propargylic alcohols in 61~93% e.e. (Table I). This catalyst was found to work for both aromatic and aliphatic acetylenes, while aliphatic aldehydes were found to give low enantioselectivities (36% ee for the alkynylation of cyclohexanecarboxaldehyde). These results compared favorably with other known amino alcohol ligands in similar reactions.

(1*R*,2*S*,3*R*)-12 R_1=Ph R_2=Ph
(1*R*,2*R*,3*S*)-13 R_1=Ph R_2=Ph
(1*S*,2*S*,3*R*)-14 R_1=Ph R_2=Ph
(1*R*,2*R*,3*S*)-15 R_1=CH_3 R_2=Ph

X-ray structure of (1R,2S,3R)-12

Table I. Asymmetric Alkynylation of Aromatic Aldehydes in the Presence of (1*R*,2*S*,3*R*)-12[a]

ArCHO + H—≡—R → ($ZnMe_2$, toluene; ligand, 10mol%) → HO(H)C*(Ar)—≡—R

Entry	*Aldehyde*	*R*	*T (℃)*	*Catalyst/ substrate*	*Conv. (%)*	*E.e. (%)*
1	Benzaldehyde	Ph	0	0.10	>95	70(-)(*S*)
2	Benzaldehyde	Ph	-20	0.10	>95	79(-)(*S*)
3	2-Chlorobenzaldehyde	Ph	0	0.01	20	3(+)
4	2-Chlorobenzaldehyde	Ph	0	0.05	45	12(+)
5	2-Chlorobenzaldehyde	Ph	0	0.10	>95	85(+)
6	2-Chlorobenzaldehyde	Ph	0	0.20	>95	88(+)
7	2-Chlorobenzaldehyde	Ph	-20	0.10	>95	89(+)
8	2-Chlorobenzaldehyde	Ph	-20	0.20	>95	93(+)
9	2-Fluorobenzaldehyde	Ph	0	0.10	>95	87(+)
10	2-Bromobenzaldehyde	Ph	0	0.10	>95	90(+)
11	2-Nitrobenzaldehyde	Ph	0	0.10	>95	87(+)
12	2-Nitrobenzaldehyde	C_3H_7	0	0.10	>95	85(+)
13	2-Anisaldehyde	Ph	0	0.10	>95	71(+)
14	2-Tolualdehyde	Ph	0	0.10	>95	71(+)
15	2-Naphthaldehyde	Ph	0	0.10	>95	61(+)

a Aldehyde: alkyne: Me_2Zn = 1: 2.4: 2.2; toluene as solvent.

The successful resolution of racemic BINOL makes optically pure BINOL an easily accessible and readily available chiral compound (*49*). Many publications on the use of BINOL in combination with metal salts in asymmetric catalysis have been cited in literature. Our previous studies (*50,51*) revealed that chiral catalysts derived from 5,5',6,6',7,7',8,8'-octahydro-1,1'-bi-2-naphthyl ligands (H_8-BINOL) exhibited higher efficiency and enantioselectivity for many asymmetric catalytic reactions than those using BINOL ligand, probably due to the steric and electronic modulations in the binaphthyl backbone. In this respect, we examined the effect of BINOL, H_8-BINOL and their derivatives in asymmetric alkynylation.

Table II. Alkynylation of Aldehydes Using Titanium Catalysts with (*R*)-H_8-BINOL Ligands[a]

RCHO + H—≡—Ph → (*R*)-H_8-BINOL + Ti(OiPr)$_4$, ZnMe$_2$ / THF → HO(*)(H)C(R)—≡—Ph

Entry	*Aldehyde*	*Yield (%)*	*E.e. (%)*	*Config.*
1	Benzaldehyde	85(84)[b]	92(90)[b]	(-)(*S*)
2	2-Chlorobenzaldehyde	90(88)[b]	76(64)[b]	(+)
3	3-Chlorobenzaldehyde	87(88)[b]	95(92)[b]	(-)
4	4-Chlorobenzaldehyde	91(87)[b]	94(92)[b]	(-)
5	4-Tolualdehyde	84(83)[b]	86(86)[b]	(-)
6	4-Fluorobenzaldehyde	82	87	(-)
7	4-Bromobenzaldehyde	89	94	(-)
8	4-Nitrobenzaldehyde	89	95	(-)
9	3-Nitrobenzaldehyde	88	96	(-)
10	2-Naphthaldehyde	75	80	(+)
11	4-Trifluoromethylbenzaldehyde	89	93	(-)
12	$(CH_3)_2CHCHO$	84	82	(-)(*S*)
13	$C_6H_{11}CHO$	86	74	(+)(*S*)
14	$CH_3CH_2CH_2CHO$	87	77	(-)

a Aldehyde: ligand: Ti(OiPr)$_4$: Me$_2$Zn = 1: 0.2: 1.5: 1.2 (molar ratio); THF as solvent; 0°C for 18 hours. b Data in brackets are from experiments using (*R*)-BINOL ligand under otherwise identical conditions.

We tried the asymmetric alkynylation using a complex generated *in situ* from titanium tetraisopropoxide and (*R*)-BINOL or (*R*)-H_8-binaphthol. The method was quite simple and the preliminary results were highly encouraging. Diverse aldehydes were converted to the corresponding propargylic alcohols with very good enantioselectivities (up to 96% e.e.) and yields (Table II). The reactions catalyzed by (*R*)-H_8-BINOL gave significantly higher e.e.s than those obtained from (*R*)-BINOL. It was observed that the *ortho*-substituted benzaldehydes gave products with lower e.e. than those from *para*- or *meta*-substituted benzaldehydes. The system was also applicable to aliphatic aldehydes and moderate to good e.e.s were obtained in most cases. The best enantioselectivity (96% e.e.) was obtained in the alkynylation of 3-nitrobenzaldehyde. To the best of our knowledge, this was the first example of the use of titanium –BINOL catalysts in asymmetric alkynylation.

In a different development, Mikami *et al.* reported the self-assembly of several chiral ligand components into a highly enantioselective titanium catalyst for carbonyl-ene reactions (*52,53*). The interesting results clearly demonstrated that both the rate and the enantioselectivity were enhanced when a combination of chiral ligand components was used instead of its single chiral ligand component in the enantioselective catalysis of the carbonyl-ene reaction. On the basis of this finding, we became interested in applying this concept of self-assembled-catalyst in asymmetric alkynylation.

Preliminary experiments showed that in combination with chiral BINOL, other chiral ligands such as diol or sulfonamide improved the catalytic activity and enantioselectivity of the alkynylation of aldehydes (Table III). Based on the observation that the use of (*R*)-BINOL in this self-assembled catalyst system produced *S* configuration of product, while the use of (*S*)-BINOL gave the product in *R* configuration (entry 7-8, 16-17), we concluded that the chiral ligand R^1 (BINOL) dominated the stereochemistry of this reaction, and the chiral ligand R^2 acted as an activator in this self-assembled catalyst system. Detailed study revealed that the ratio of chiral ligands (R^1 + R^2) to $Ti(OiPr)_4$ was important in influencing the enantioselectivity of the catalyst and the optimal ratio was around 1: 1.25 – 1.5.

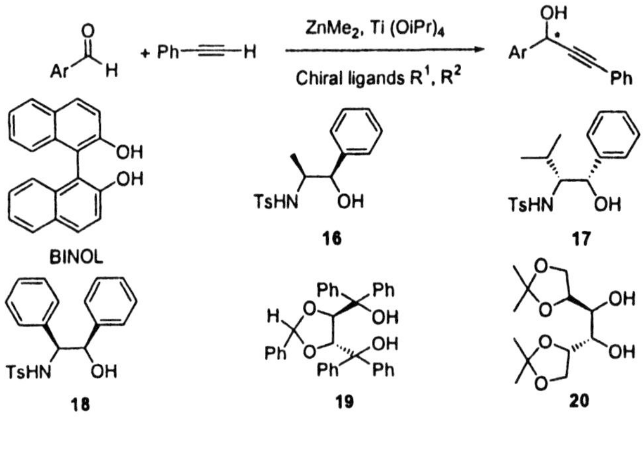

$ZnMe_2$, Ti $(OiPr)_4$
Chiral ligands R^1, R^2

BINOL 16 17 18 19 20

Table III. Asymmetric Alkynylation of Benzaldehyde in the Presence of Activator[a]

Entry	R^1	R^2	$Ti/(R^1+R^2)$	*Yield (%)*	*E.e. (%)*	*Config.*
1	none	**16**	7/1	100	0	-
2	none	**17**	7/1	100	0	-
3	none	**18**	7/1	100	0	-
4	none	**19**[b]	7/1	10[b]	0	-
5	none	**20**[b]	7/1	12[b]	0	-
6	(*R*)-BINOL	none	7/1	90	77	*S*
7	(*R*)-BINOL	**16**	7/1	90	87	*S*
8	(*S*)-BINOL	**16**	7/1	88	90	*R*
9	*rac*-BINOL	**16**[c]	7/1	91	18	*R*
10	(*S*)-BINOL	**17**	7/1	89	89	*R*
11	(*S*)-BINOL	**18**	7/1	90	90	*R*
12	(*S*)-BINOL	**19**	7/1	87	86	*R*
13	(*S*)-BINOL	**20**	7/1	86	85	*R*
14	(*S*)-BINOL	**16**	3.5/1	85	93	*R*
15	none	**16**	1.5/1	43	0	-
16	(*S*)-BINOL	**16**	1.5/1	83	96	*R*
17	(*R*)-BINOL	**16**	1.5/1	80	94	*S*
18	(*S*)-BINOL	**16**	1.25/1	81	**97**	*R*
19	(*S*)-BINOL	none	1.25/1	74	91	*R*
20	(*S*)-BINOL	**16**	1/1	84	94	*R*
21	(*S*)-BINOL	none	1/1	66	87	*R*

a Aldehyde: $ZnMe_2$: R^1: R^2 =1: 2: 0.1: 0.1 (molar ratio); the reaction was carried out at 0°C under nitrogen atmosphere for 24-48 hours. b The major product was α-methylbenzyl alcohol. c 5 % R^2 ligand.

The results of the addition of alkynes to a variety of aromatic aldehydes were summarized in Table IV. Good to excellent enantiomeric excesses were obtained (88~>99% e.e.). These results are commensurate with the highest enantioselectivities of the products reported for the reactions of terminal alkynes with aromatic aldehydes catalyzed by stoichiometric amount of *N*-methylephedrine and $Zn(OTf)_2$.

In conclusion, this study showed that a combination of chiral ligands, such as BINOL and sulfonamide, together with $Ti(OiPr)_4$ generated a highly enantioselective catalyst for the synthesis of propargylic alcohols in up to >99% e.e. Further study of the mechanism of the reaction and the application of this catalyst system for other asymmetric catalytic reactions are in progress.

Table IV. Alkynylation of Aromatic Aldehydes Catalyzed by the Combination of (*S*)-BINOL and Sulfonamide 16[a]

Entry	*Aldehyde*	*Yield (%)*	*E.e. (%)*
1	Benzaldehyde	83	96
2	2-Nitrobenzaldehyde	83	88
3	3-Nitrobenzaldehyde	82	**99.7**
4	4-Nitrobenzaldehyde	82	99
5	4-Bromobenzaldehyde	85	99
6	3-Chlorobenzaldehyde	84	97
7	4-Chlorobenzaldehyde	86	95
8	2-Naphthaldehyde	81	95
9	4-Anisaldehyde	78	95
10	4-Tolualdehyde	79	92

a The ratio of ligands (BINOL + sulfonamide **16**) to titanium tetraisopropoxide was 1.0: 1.5, 0°C, 24-48h.

Acknowledgements

We thank the Hong Kong Polytechnic University ASD Fund, the Hong Kong Research Grants Council (Project No. ERB03) and the University Grants Committee Areas of Excellence Fund (AoE P/10-01) for financial support of this study.

References

1. Marshall, J. A.; Wang, X. J. *J. Org. Chem.* **1992**, *57*, 1242-1252.
2. Roush, W. R; Sciotti, R. J. *J. Am. Chem. Soc.* **1994**, *116*, 6457-6458.
3. Myers, A. G.; Zheng, B. *J. Am. Chem. Soc.* **1996**, *118*, 4492-4493.
4. Brinkmeyer, R. S.; Kapoor, V. M. *J. Am. Chem. Soc.* **1977**, *95*, 8339-8341.
5. Noyori, R.; Tomino, I.; Yamada, M.; Nishizawa, M. *J. Am. Chem. Soc.* **1984**, *106*, 6717-6725.
6. Midland, M. M.; Tramontano, A.; Kazubski, A.; Graham, R. S.; Tsai, D. J. S.; Cardin, D. B. *Tetrahedron,* **1984**, *40*, 1371-1380.
7. Ramachandran, P. V.; Teodorovic, A. V.; Rangaishenvi, M. V.; Brown, H. C. *J. Org. Chem.* **1992**, *57*, 2379-2386.
8. Bach, J.; Berenguer, R.; Garcia, J.; Loscertales, T.; Vilarrasa, J. *J. Org.Chem.* **1996**, *61*, 9021-9025.
9. Ishihara, K.; Mori, A.; Arai, I.; Yamamoto, H. *Tetrahedron Lett.* **1986**, *27*, 983-986.
10. Bradshaw, C. W.; Hummel, W.; Wong, C. H. *J. Org. Chem.* **1992**, *57*, 1532-1536.
11. Ansari, M. H.; Kusumoto, T.; Hiyama, T. *Tetrahedron Lett.* **1993**, *34*, 8271-8274.
12. Burgess, K.; Jennings, L. D. *J. Am. Chem. Soc.* **1991**, *113*, 6129-6139.
13. Jeromin, G. E.; Scheidt, A. *Tetrahedron Lett.* **1991**, *32*, 7021-7024.
14. Schubert, T.; Hummel, W.; Kula, M. R.; Muller, M. *Eur. J. Org. Chem.* **2001**, *22*, 4181-4187.
15. Schubert, T.; Hummel, W.; Muller, M. *Angew. Chem. Int. Ed. Engl.* **2002**, *41*, 634-637.
16. Corey, E. J.; Helal, C. J. *Tetrahedron Lett.* **1995**, *36*, 9153-9156.
17. Helal, C. J.; Magriotis, P. A.; Corey, E. J. *J. Am. Chem. Soc.* **1996**, *118*, 10938-10939.

18. Parker, K. A.; Ledeboer, M. W. *J. Org. Chem.* **1996**, *61*, 3214-3217.
19. Matsumura, K.; Hashiguchi, S.; Ikariya, T.; Noyori, R. *J. Am. Chem. Soc.* **1997**, *119*, 8738-8739.
20. Oppolzer, W.; Radinov, R. N. *J. Am. Chem. Soc.* **1993**, *115*, 1593-1594.
21. Noyori, R.; Kitamura, M. *Angew. Chem. Int. Ed. Engl.* **1991**, *30*, 49-69.
22. Williams, D. R; Fromhold, M. G. *Synlett,* **1997**, 523-524.
23. Pu, L.; Yu, H. B. *Chem. Rev.* **2001**, *101*, 757-824.
24. Mukaiyama, T.; Suzuki, K.; Soai, K.; Sato, T. *Chem. Lett.* **1979**, 447-448.
25. Mukaiyama, T.; Suzuki, K. *Chem. Lett.* **1980**, 255-256.
26. Yamamoto, Y.; Nishii, S.; Maruyama, K. *J. Chem. Soc. Chem. Commun.* **1986**, 102-103.
27. Krause, N.; Seebach, D. *Chem. Ber.* **1987**, *120,* 1845-1851.
28. Niwa, S.; Soai, K. *J. Chem. Soc. Perkin Trans.1,* **1990**, *4*, 937-943.
29. Soai, K.; Niwa, S. *Chem. Rev.* **1992**, *92*, 833-856.
30. Tombo, G. M. R.; Didier, E.; Loubinoux, B. *Synlett,* **1990**, 547-548.
31. Chelucci, G.; Conti, S.; Falorni, M.; Giacomelli, G. *Tetrahedron*, **1991**, *47*, 8251-8258.
32. Conti, S.; Falorni, M.; Giacomelli, G.; Soccolini, F. *Tetrahedron*, **1992**, *48*, 8993-9000.
33. Yamaguchi, M.; Hayashi, A.; Hirama, M. *Chem. Lett.* **1992**, 2479-2482.
34. Baldoli, C.; Buttero, P. D.; Licandro, E.; Maiorana, S.; Papagni, A.; Torchio, M. *Tetrahedron Lett.* **1993**, *34,* 7943-7946.
35. Corey, E. J.; Cimprich, K. A. *J. Am. Chem. Soc.* **1994**, *116*, 3151-3152.
36. Ishizaki, M.; Hoshino, O. *Tetrahedron: Asymmetry*, **1994**, *5*, 1901-1904.
37. Li, Z.; Upadhyay, V.; DeCamp, A. E.; DiMichele, L.; Reider, P. *Synthesis*, **1999**, 1453-1458.
38. Boyall, D.; Lopez, F.; Sasaki, H.; Frantz, D.; Carreira, E. M. *Org. Lett.* **2000**, *2*, 4233-4236.
39. Frantz, D. E.; Fassler, R.; Carreira, E. M. *J. Am. Chem. Soc.* **2000**, *122*, 1806-1807.
40. Frantz, D. E.; Fassler, R.; Tomooka, C. S.; Carreira, E. M. *Acc. Chem. Res.* **2000**, *33*, 373-381.
41. Anand, N. K.; Carreira, E. M. *J. Am. Chem. Soc.* **2001**, *123*, 9687-9688.
42. Boyall, D.; Frantz, D. E.; Carreira, E. M. *Org. Lett.* **2002**, *4*, 2605-2606.
43. Sasaki, H.; Boyall, D.; Carreira, E. M. *Helv. Chim. Acta*, **2001**, *84,* 964-971.
44. Moore, D.; Pu, L. *Org. Lett.* **2002**, *4*, 1855-1857.
45. Jiang, B.; Chen, Z. L.; Xiong, W. N. *J. Chem. Soc. Chem. Commun.* **2002**, 1524-1525.
46. Lu, G.; Li, X. S.; Zhou, Z. Y.; Chan, W. L.; Chan, A. S. C. *Tetrahedron: Asymmetry*, **2001**, *12,* 2147-2152.
47. Lu, G.; Li, X. S.; Chan, W. L.; Chan, A. S. C. *J. Chem. Soc. Chem. Commun.* **2002**, 172-172.
48. Li, X. S.; Lu, G.; Kwok, W. H.; Chan, A. S. C. *J. Am. Chem. Soc.* **2002**, in press.
49. Cai, D.; Hughes, D. L.; Verhoeven, T. R.; Reider, P. J. *Tetrahedron Lett.* **1995**, *36*, 7991-7994.
50. Zhang, F. Y.; Chan, A. S. C. *Tetrahedron: Asymmetry*, **1997**, *8*, 3651-3655..
51. Chan, A. S. C.; Zhang, F. Y.; Yip, C. W. *J. Am. Chem. Soc.* **1997**, *119*, 4080-4081.
52. Mikami, K.; Matsukawa, S.; Volk, T.;Terada, M. *Angew. Chem. Int. Ed. Engl.* **1997**, *36*, 2768-2771.
53. Mikami, K.; Terada, M.; Korenaga, T.; Matsumoto, Y.; Ueki, M.; Angelaud, R. *Angew. Chem. Int. Ed. Engl.* **2000**, 39, 3532-3556.

Chapter 8

An Affordable Catalyst for the Production of Amino Acids

David Ager[1], Michel van den Berg[2], Adriaan J. Minnaard[2], Ben L. Feringa[2], André H. M. de Vries[3], Charlotte E. Willans[3], Jeroen A. F. Boogers[3], and Johannes G. de Vries[2,3]

[1]DSM Pharmaceuticals, Inc., P.O. Box 1887, Greenville, NC 27835
[2]University of Groningen, Stratingh Institute, Nijenborgh 4, 9747 AG Groningen, The Netherlands
[3]DSM Research, Life Science-Advanced Synthesis and Catalysis, P.O. Box 18, 6160 MD Geleen, The Netherlands

Asymmetric hydrogenation has been demonstrated to be a useful synthetic approach to amino acids and derivatives. The asymmetric reduction step itself is usually straightforward, but the synthesis of the ligand that invokes the asymmetric induction can be difficult. Use of MonoPhos™, a phosphoramidite, allows for a low cost solution as it is derived in a single step from BINOL, an item of commerce available as either antipode. Rhodium catalyzed asymmetric hydrogenations proceed with high enantioselectivities, especially to substituted phenylalanine derivatives. A large solvent effect is observed.

Introduction

Many ligands have been developed and advocated for the asymmetric reduction of enamides by a transition metal catalyst to α-amino acid derivatives (Scheme 1) (*1,2*). Most known ligands are bidentate. The ligand's chirality can be derived from stereogenic phosphorus atoms, as in the DIPAMP (**1**) system of Knowles, or from carbon stereocenters. Examples are given in Figure 1. There has also been a number of systems developed, of which BINAP (**2**) is the best known, where the asymmetry is derived from hindered rotation.

Scheme 1.

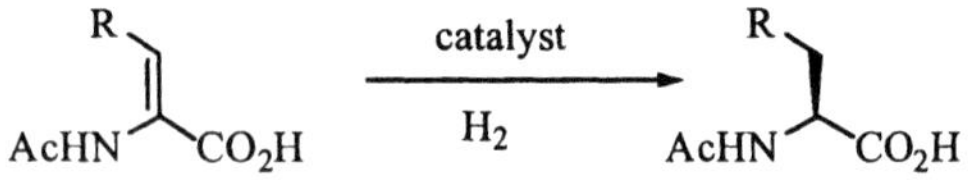

R AcHN CO_2H —catalyst, H_2→ R AcHN CO_2H

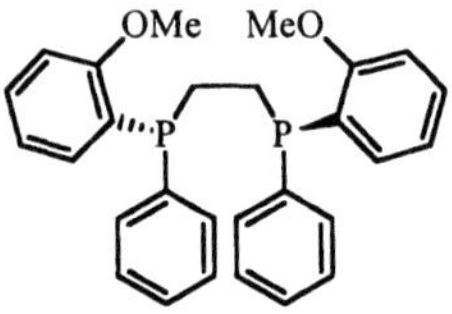

1 DIPAMP

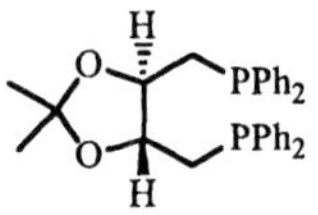

DIOP

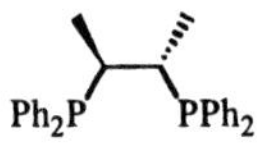

Chiraphos

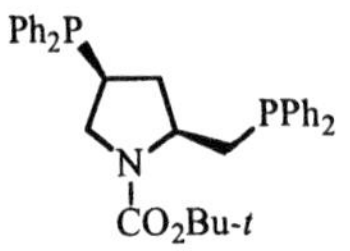

BPPM

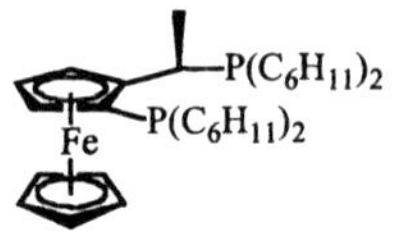

Josiphos

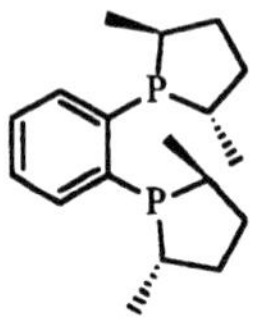

Duphos

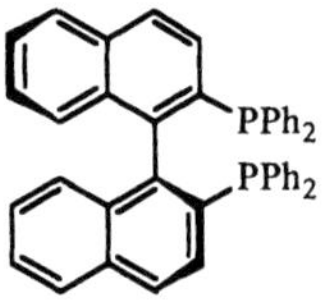

2 BINAP

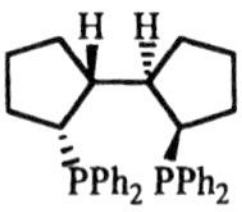

3 BICP

Figure 1: Some examples of ligands used to effect asymmetric reductions.

Whatever the source of chirality, all these systems have a major drawback in that their synthesis is expensive, can involve a long sequence, be low yielding, or tedious. Often a resolution has to be performed. In a few cases, as in BICP (**3**), the chirality originates from the chiral pool and access to the antipode can be problematic (*3*). In most papers on asymmetric hydrogenations, these problems are treated in a cursory manner or completely ignored. To perform an asymmetric hydrogenation reaction at scale, the ligand synthesis can be the greatest problem that has to be overcome. In addition, the emphasis on enantioselectivity is usually to the detriment of low catalyst usage, which can have a large impact on the economics of the process. The kinetics of the reaction are often ignored in publications, and these can also have a large effect on the economics of the process.

Although Knowles' catalyst has been known for almost thirty years, and has been used at industrial scale (*2*), there are still relatively few asymmetric hydrogenations performed at scale. If we consider Knowles' catalyst, which has been used for the production of L-Dopa (Scheme 1; R = 3-AcO,4-MeOC$_6$H$_3$; cat = [Rh(DIPAMP)(COD)]BF$_4^-$), the synthesis of the ligand is long, low yielding (*4*) and tedious (*5*). However, a wide range of enamides can be reduced and the method is useful for the production of unnatural amino acids as long as there is no substituent at the β-position (*4*). The development of the DuPhos and related systems overcame the β-substituent problem, but again the synthesis of the ligand is not a trivial undertaking (*6*). There is also a dogma that bidentate ligands are required to obtain high asymmetric induction.

A recent survey by Blaser *et al.* shows that at best a dozen processes based on asymmetric hydrogenation have been implemented for the production of enantiopure intermediates on ton-scale (*7*). We believe that this scant use of such a well-studied technology is caused by the following main reasons: Time-to-market constraints often do not allow sufficient time to develop a catalytic process for new products. The costs of metal or ligand may be too high. The third reason can be poor availability of ligands (both lab and production scale) on short notice. The result of these problems in the fine chemical industry, especially for pharmaceutically related products, is that methodology that is easier to screen and implement is used in preference to asymmetric hydrogenations.

MonoPhos™ reductions

With the knowledge that there is still a need for better asymmetric hydrogenation catalysts, it was fascinating that phosphoramidite derivatives of binaphthol (**4**) (BINOL) have been used to effect asymmetric conjugate additions to enones (*8*). The ligand system, was therefore, tried in the context of an asymmetric hydrogenation as libraries of ligands could be prepared by a simple procedure (*vide infra*). The common perception that a bidentate ligand is required for good asymmetric induction was found not to be the case. In addtion to the synthesis of α-amino acids, application of monodentate phosphoramidites in the rhodium catalysed hydrogenations of aromatic enamides (*9*) and dehydro-beta-amino acids (*10*) have recently been reported.

The initial experiments to prepare α-amino acids were performed with the dimethylamino phosphoramidite, that has become known as MonoPhos™ (**5**). The model enamide is methyl α-*N*-acetylamidocinnamate (**6**) (Scheme 2).

Scheme 2.

CO_2Me NHAc **6** — $[Rh(COD)_2]^+BF_4^-$, ligand; H_2, solvent → CO_2Me NHAc

Ligand = MonoPhos™

P–NMe_2

5

The initial findings were that there is a strong solvent effect. Use of methanol provides moderate asymmetric induction but when methylene chloride is the solvent, excellent asymmetric induction is observed (Scheme 2 and Table I) (*11,12*). The reaction also works well for a variety of substituted phenylalanine derivatives (Table II and Scheme 3) (*13*).

The initial hydrogenation experiments were performed in Schlenk tubes at 1 bar using a small magnetic stirring bar. Rates of these reactions were not very high. However, upon transferring the hydrogenation into well-stirred autoclaves, and raising the pressure to 5 bar we could raise the turnover frequency (TOF) to 300 h^{-1} (Table II, Entry 3). At high pressure (60 bar) an unoptimized TOF of 1667 h^{-1} has been achieved. Obviously, the reaction has a positive order in hydrogen pressure. Most gratifyingly we found that, unlike with many bisphosphine ligands, the enantioselectivity does not decrease with increasing pressure.

A variety of substituents on the phenyl group in the substrate have little or no effect on the enantioselectivity of the hydrogenation, although the rate was affected significantly (Table II).

Table I. Summary of enantioselectivity for the preparation of phenylalanine from 6 as described in Scheme 2 with 1 bar of hydrogen.

Solvent	*Temp*	*ee (%)*
CH_3OH	rt[a]	70
CH_2Cl_2	rt	95
CH_2Cl_2	5°C	97
THF	rt	93
Me_2CO	rt	92
EtOAc	rt	95
$PrOCH_2CH_2OH$	rt	77

[a] Denotes room temperature.

Scheme 3.

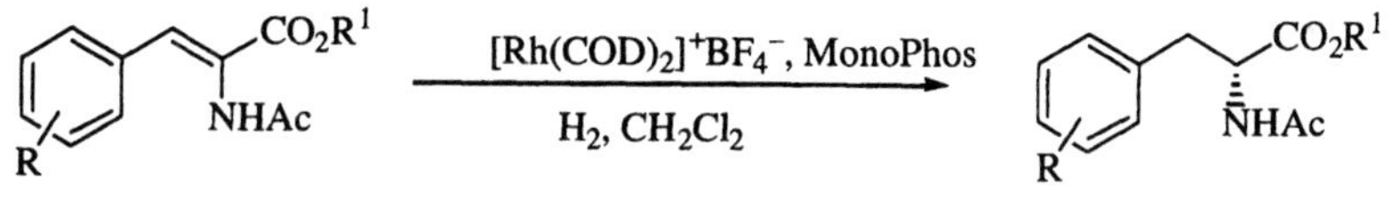

CO_2R^1, NHAc, R → $[Rh(COD)_2]^+BF_4^-$, MonoPhos; H_2, CH_2Cl_2 → CO_2R^1, NHAc, R

Table II. Reductions of substituted phenylalanine derivatives as shown in Scheme 3.[13]

Entry	*R*	*R^1*	*Solvent*	*pH_2 (bar)*	*Time*	*ee (%)*
1	H	H	CH_2Cl_2	5	3 h	97
2	H	Me	CH_2Cl_2	1	3 h	95
3	H	Me	CH_2Cl_2	5	40 min	95
4	H	Me	EtOAc	1	2 h	96
5	3-MeO	Me	CH_2Cl_2	5	2 h	97
6	4-MeO	Me	CH_2Cl_2	5	2 h	94
7	4-AcO, 3-MeO	Me	EtOAc	5	nd[a]	96
8	4-F	H	CH_2Cl_2	27	10 min	93
9	4-F	Me	CH_2Cl_2	5	25 min	96
10	3-F	H	CH_2Cl_2	10	2 h	96
11	3-F	Me	CH_2Cl_2	5	30 min	95
12	2-F	Me	CH_2Cl_2	5	15 min	95
13	4-Cl	Me	CH_2Cl_2	5	20 min	94
14	3,4-Cl_2	H	CH_2Cl_2	5	2 h	97
15	3,4-Cl_2	Me	CH_2Cl_2	5	30 min	99
16	3-NO_2	Me	CH_2Cl_2	5	2 h	95
17	4-NO_2	Me	CH_2Cl_2	5	nd	95
18	4-F, 3-NO_2	Me	CH_2Cl_2	5	2 h	95
19	4-Ph	Me	CH_2Cl_2	5	25 min	95
20	3-F, 4-Ph	Me	CH_2Cl_2	5	25 min	93
21	4-Ac	Me	CH_2Cl_2	5	15 min	99
22	4-PhCO	Me	CH_2Cl_2	5	30 min	94
23	4-CN	Me	CH_2Cl_2	5	18 h	92
24	1-naphthyl[b]	Me	CH_2Cl_2	5	10 min	93

[a] Denotes not determined. [b]This is the aryl substituent.

Ligand synthesis

Although MonoPhos™ is a monodentate ligand, good asymmetric induction is observed. The power of this approach is that the ligand is readily available from BINOL (**4**) (Scheme 4) (*14*). Other analogues **7** of MonoPhos™ are available by substitution of the amine moiety or by a short synthesis (Scheme 4) (*15-17*).

Scheme 4.

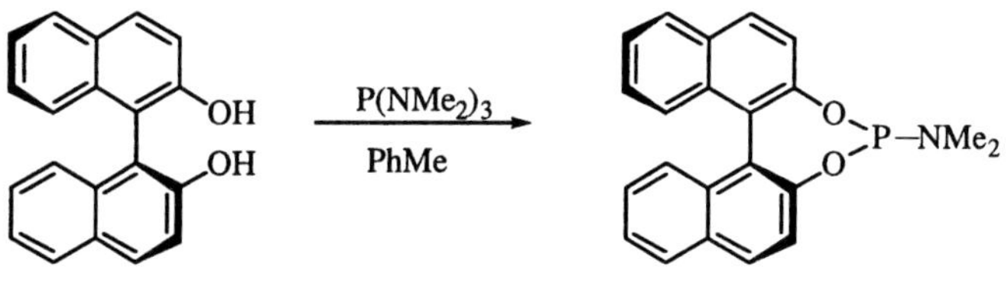

$P(NMe_2)_3$ / PhMe: **4** → **5**

Scheme 5.

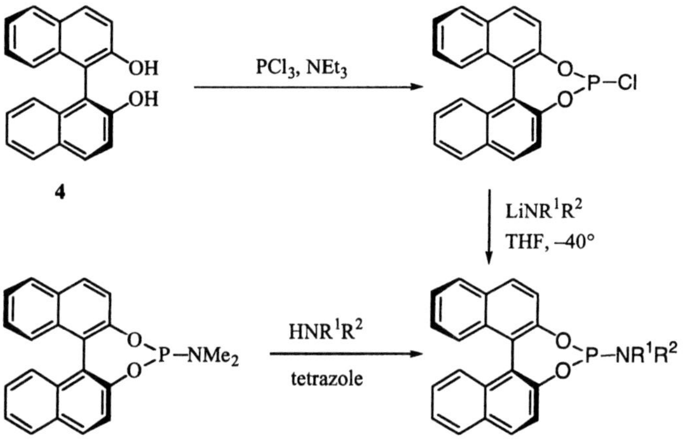

PCl_3, NEt_3: **4** → chlorophosphite; $LiNR^1R^2$, THF, –40° → **7**

HNR^1R^2, tetrazole: **5** → **7**

The complex formation between the ligand and the metal can be performed prior to the reduction either as a separate step, or *in situ*. The procedure alleviates some of the problems associated with oxygen sensitivity seen with other systems.

Mechanistic considerations

This aspect of the work is still ongoing. Many studies have appeared investigating the mechanism of rhodium-bisphosphine catalyzed asymmetric hydrogenation of dehydroamino acids (*18*). In the Halpern mechanism oxidative addition of hydrogen to the Rh-bisphosphine olefin complex is the rate-determining step (*19,20*). Recently, Imamoto has published evidence for the formation of a Rh-bisphosphine dihydride complex at low temperature, which has reopened the debate on the mechanism (*21*).

One of the more surprising characteristics of the use of monodentate phosphoramidites in rhodium-catalyzed hydrogenations is the effect of the ligand/rhodium ratio. Phosphoramidites bind very strongly to rhodium, in spite of their reduced sigma donating properties due to their greatly enhanced π acceptor properties as compared to phosphines. It thus did not come as a surprise that catalytic activity ceased when the L/Rh ratio was increased to 3 (Figure 2) in the hydrogenation of **6**.

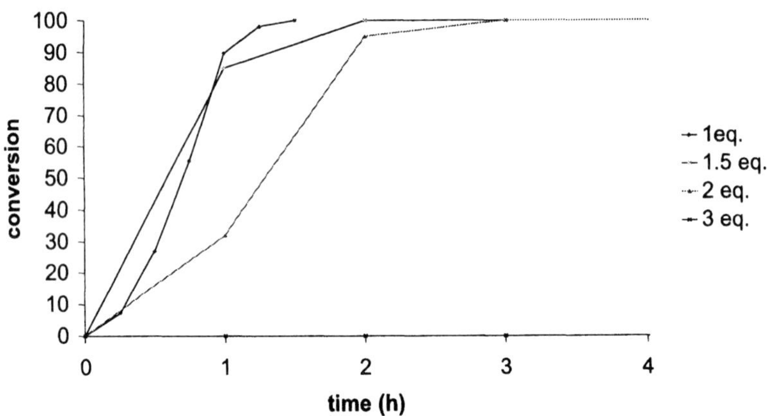

Figure 2: Rate dependence on the MonoPhos:Rh ratio for the reduction of ***6***

However, when the ratio of **5**:Rh was lowered to 1.5 or 1 we found an unexpected increase of the rate. Even more interesting was the finding that the ee remained the same over the whole L/Rh range from 1-2 equivalents of MonoPhos:Rh and remained the same throughout the hydrogenation reaction. This suggests that a single rhodium species is the active catalyst in all these hydrogenations independent of the **5**:Rh ratio. We have performed these experiments both at 1 bar with 5 mol% of catalyst and at 15 bar with 0.015 mol% of catalyst and found essentially the same results. Thus far we had assumed that in the entire catalytic cycle two ligands would be bound to rhodium. However, these results suggest that this may not be the case.

To shed more light on this dilemma, we decided to perform a test for non-linear effects after the work of Kagan *et al* (*22*). We performed this test using MonoPhos™ (**5**) of varying degrees of enantiomeric purity in the rhodium-catalyzed hydrogenation of **6**. As can be seen from Figure 3, we obtained a weak, but reproducible, positive non-linear effect. Although this definitely establishes the presence of complexes with two or more ligands it still does not rule out the possibility that these complexes dissociate into a mono-ligated complex, which could be the actual catalytic species . It is clear that more information is necessary.

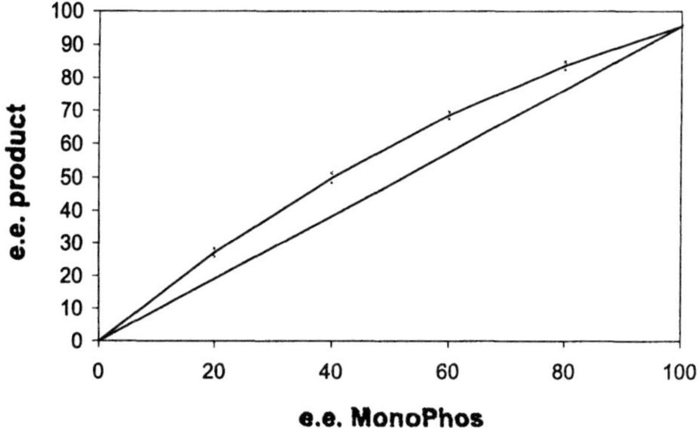

Figure 3: Asymmetric amplification with MonoPhos.

As a comparison a number of different ligands were prepared. One variation was changing of the nitrogen substituents (**8**,**9**). Hydrogenation of BINOL over PtO_2 (*23*), followed by reaction with HMPT gave **10**. The effect of substituents on the BINOL skeleton in the 6- (**11**) and 3-positions (**12**) was probed. In addition, bridged systems with two phosphoramidite moieties were prepared (**13-15**).

8 **9** **10**

11 **12**

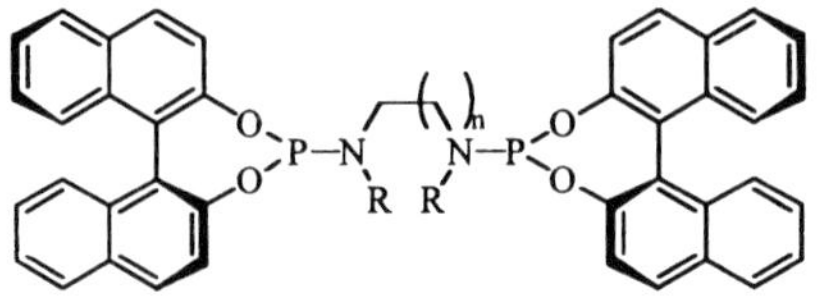

13; n = 1, R = Me
14; n = 2, R = *i*-Pr
15; n = 2, R = (*S*)-PhMeCH

The ligands were tested in the Rh-catalysed hydrogenation of a limited number of dehydrophenylalanine derivatives; most experiments were performed on **6**. In the dehydroamino acid hydrogenations the two methyl substituents on nitrogen as in **5** are unsurpassed up to now. The 6,6'-dibromo-MonoPhos **11** behaved very similar to MonoPhos in the hydrogenation of **6**, although the rate is somewhat lower (Figure 4). Enantioselectivity did suffer when this ligand was applied for the hydrogenation of the 4-fluoro-derivative. The effect of the 3,3'-dimethyl substituents of **12** was more marked and led to a strongly reduced rate as well as a somewhat lower enantioselectivity in the hydrogenation of **6**. Surprisingly, the configuration of the product is the same as that of the BINOL, in contrast to results with the other BINOL ligands. Hydrogenations with octahydro-MonoPhos **11** gave results that were comparable to those obtained with MonoPhos itself. This was recently confirmed by the work of Chan *et al.* (*24*). Use of bidentate ligands such as **13-15** was not so successful. In methanol hardly any reaction took place, although in CH_2Cl_2 a reasonable rate could be attained, depending on bridge length. Nevertheless, enantioselectivity was much lower than with the monodentate ligands.

A comparison of the kinetics of the hydrogenation of **6** using ligands **5, 10-12** is given in Figure 4. These traces represent hydrogen uptake during hydrogenation at 5 bar pressure that were performed simultaneously in the Endeavor™, a semi-automated device which allows 8 parallel gas-liquid reactions with continuous monitoring of hydrogen uptake. The first 30 minutes are used for repeated purging with N_2. After the solution has been pressurised with H_2 the reaction starts almost immediately and seemingly follows zero order kinetics with **5** and **10**. Although hydrogenation using **11** is clearly slower the kinetic profile is very similar to that of **5** and **10**. Quite different kinetics are observed with the 3,3'-dimethyl derivative **12**; an induction period is apparent and after a much slower reaction the rate gradually decreases which might point to some catalyst deactivation. The induction period usually is related to the catalyst activation step by hydrogenation of the remaining bound COD, which can be very slow with some ligands (*25-27*). However, a similar profile would be obtained if the reaction was first order in substrate. At present, we have no kinetic data to support this theory.

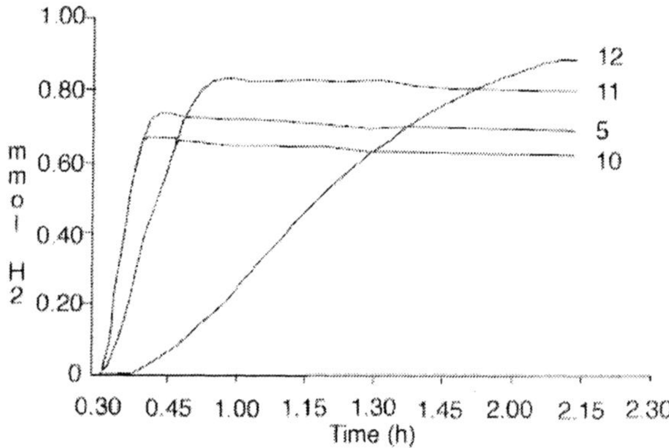

*Figure 4: Hydrogen uptake in the rhodium catalysed hydrogenation of **6** at 5 bar with different ligands.*

So far we have not been able to glean much information from NMR experiments as broad absorptions were observed. We thus decided to use electrospray mass spectrometry to further investigate the rhodium species that are present during hydrogenation. Samples of a hydrogenation of **6** in CH_2Cl_2 using 5 mol% of $[Rh(nbd)_2]BF_4$/MonoPhos (**5**) (1:2) as catalyst at 1 bar H_2 pressure were taken at regular intervals and examined by ES-MS (Table III).

Table III. Rhodium species observed with ES-MS in the hydrogenation of 6 with $Rh(nbd)_2BF_4$ /5.

Time (min)	*Species*[a]
30	RhLS*, RhL_2 (nbd), RhL_2S*, RhL_3, RhL_3S
60	RhLS*, RhL_2(nbd), RhL_2S*, RhL_3, RhL_4
120	RhLS*, RhL_2S*, RhL_3, RhL_4

* Small peaks. [a] Where L= ligand, S=substrate.

As can be seen from Table III the catalyst containing two ligands and 1 norbornadiene (nbd) remains present in solution for at least 60 min; after 120 min it has disappeared completely. Its persistent presence might well be due to the poor mass-transfer in these magnetically stirred solutions. We have never experienced a lag-time in the hydrogenations that were performed at higher pressures in the autoclave. We also observe complexes containing 1, 2 or 3 ligands and 1 substrate molecule, however these peaks are very small. The dependence of the rate on the hydrogen pressure confirms the fact that oxidative addition of hydrogen on the RhL_n(Substrate) complex is the rate-determining step as in Halpern's original proposal (*19*). The most important conclusion of this experiment is that part of the rhodium is locked up in complexes such as RhL_3 or RhL_4 that would not seem to be part of the catalytic cycle. The higher rate of hydrogenation when the L/Rh ratio is less than 2 might well be explained by a shift in this unfavorable equilibrium towards the lower ligated species. That these higher ligated rhodium species are present in substantial amounts was proven by the isolation of $[Rh(\mathbf{5})_4]BF_4$ as pentane/CH_2Cl_2 solvate crystals from one of the 1 bar hydrogenation experiments with L/Rh 2.2. The crystal structure is shown in Figure 5. The highly symmetrical structure of the square planar complex is remarkable. Surprisingly, the dimethylamino groups of all 4 ligands are located in the same hemisphere. The helical structure of one of the ligands is shown in Figure 6. We are currently performing kinetic experiments, which should allow us to determine the composition of the catalytic species.

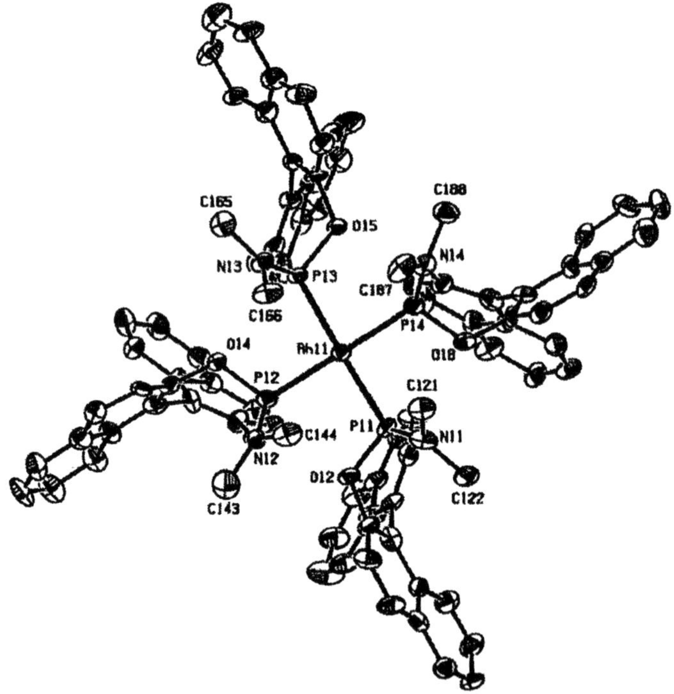

Figure 5: ORTEP representation of $[Rh(5)_4]BF_4$; hydrogen atoms are omitted for clarity.

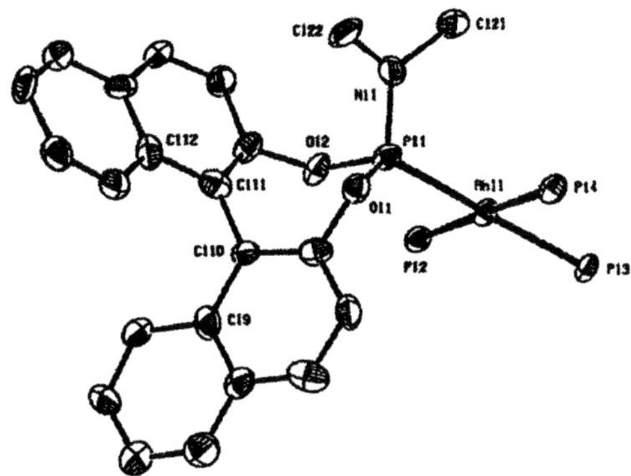

Figure 6: Detail of ORTEP representation of complex [Rh(5)$_4$]BF$_4$ showing one of the ligands with its helical shaped diol moiety.

Summary

Phosphoramidites are excellent ligands for rhodium-catalyzed asymmetric hydrogenations. MonoPhos™ (**5**) provides the highest enantioselectivity in most cases for the synthesis of α-amino acid derivatives. Although the exact nature of the catalytic species is not known, likely candidates are complexes with one or two phosphoramidite ligands attached to rhodium. Work continues to understand the interactions between the ligands and metal. The reactions have been scaled up.

References

1. *Comprehensive Asymmetric Catalysis*, Eds: Jacobsen, E. N.; Pfaltz, A.; Yamamoto, H. Springer, Berlin, **1999.**
2. Laneman, S. A. in *Handbook of Chiral Chemicals*, Ager, D. J. Ed.; Marcel Dekker, New York, **1999**, p143.
3. Zhu, G.; Cao, P.; Zhang, X. *J. Am. Chem. Soc.*, **1997**, *119*, 1799.
4. Koenig, K. E. in *Asymmetric Synthesis;* Morrison, J. D. Ed.; Academic Press, Inc.: Orlando, FL, **1985**; *Vol. 5*; p. 71.
5. Vineyard, B. D.; Knowles, W. S.; Sabacky, M. J.; Bachman, G. L.; Weinkauff, D. J. *J. Am. Chem. Soc.* **1977,** *99*, 5946.

6. Burk, M. in *Handbook of Chiral Chemicals*, Ager, D. J. Ed.; Marcel Dekker, New York, **1999**, p339.
7. Blaser, H. U.; Spindler, F.; Studer, M. *Appl. Catal. A: General* **2001**, *221*, 119.
8. Feringa, B. L. *Acc. Chem. Res.* **2000**, *33*, 346.
9. van den Berg, M.; Haak, R.M.; Minnaard, A.J.; de Vries, A.H.M.; de Vries, J.G.; Feringa, B.L. *Adv. Synth. Catal.*, **2002**, *344*, 1003.
10. Peña, D.; Minnaard, A.J.; de Vries, J.G.; Feringa, B.L. *J. Am. Chem. Soc.*accepted for publication.
11. van den Berg, M; Minnaard, A.J.; Schudde, E.P; van Esch, J.; de Vries, A.H.M.; de Vries, J.G.; Feringa, B.L.; *J. Am. Chem. Soc.* **2000**, *122*, 11539.
12. van den Berg, M.; Minnaard, A.J.; de Vries, J.G.; Feringa, B.L. (DSM N.V.), World Patent WO 02/04466, **2002**.
13. van den Berg, M.; Minnaard, A.J.; Haak, R.M.; Leeman, M.; Schudde, E.P.; Meetsma, A.; Feringa, B.L.; de Vries, A.H.M.; Maljaars, C.E.P.; Willans, C.E.; Hyett, D.; Boogers, J.A.F.; Henderickx, H.J.W.; de Vries, J.G. *Adv. Synth. Catal.*, accepted for publication.
14. de Vries, A. H. M.; Meetsma, A.; Feringa, B. L. *Angew. Chem Int. Ed,* **1996**, *35*, 2374.
15. Feringa, B.L.; Naasz, R; Imbos, R. Arnold, L.A. in *Modern Organocopper Chemistry,* N. Krause, Ed., Wiley-VCH, 2002, Chapter 7.
16. van Rooy, A.; Burgers, D.; Kamer, P. C. J.; van Leeuwen, P. W. N. *Recl. Trav. Chim. Pays-Bas* **1996**, *115*, 492.
17. Choi, Y. H.; Choi, J. Y.; Yang, H. Y.; Kim, Y. H. *Tetrahedron: Asymmetry*, **2002**, *13*, 801.
18. Brown, J. M. in *Comprehensive Asymmetric Catalysis*, Eds: Jacobsen, E. N.; Pfaltz, A.; Yamamoto, H. Springer, Berlin, **1999**, Vol 1, p 121-182.
19. Chan, A. S. C.; Pluth, J. J.; Halpern, J. *J. Am. Chem. Soc.* **1980**, *102*, 5952.
20. Brown, J. M.; Chaloner, P. A. *J. Chem. Soc., Chem. Commun.* **1980**, 344.
21. Gridnev, I. D.; Higashi, N.; Asakura, K.; Imamoto, T. *J. Am. Chem. Soc.* **2000**, *122*, 7183.
22. Kagan, H. B.; Luukas, T. O. in *Comprehensive Asymmetric Catalysis*, Eds: Jacobsen, E. N.; Pfaltz, A.; Yamamoto, H. Springer, Berlin, **1999**, *Vol 1*, p101.
23. Cram, D. J.; Helgeson, R. C.; Peacock, S. C.; Kaplan, L. J.; Domeier, L. A.; Moreau, P.; Koga, K.; Mayer, J. M.; Chao, Y.; Siegel, M. G.; Hoffmann, D. H.; Soga, G. D. Y. *J. Org. Chem.* **1978**, *43*, 1930.

24. Zeng, Q.; Liu, H.; Cui, X.; Mi, A.; Jiang, Y.; Li, X.; Choi M. C. K.; Chan, A. S. C. *Tetrahedron: Asymmetry* **2002**, *13*, 115.
25. Drexler, H.-J.; Baumann, W.; Spannenberg, A.; Fischer, C.; Heller, D. *J. Organometal. Chem.* **2001**, *621*, 89.
26. Heller, D.; Borns, S.; Baumann, W.; Selke, R. *Chem. Ber.* **1996**, *129*, 85.
27. Heller, D.; Kortus, K.; Selke, R. *Liebigs Ann.* **1995**, 575.

Chapter 9

Mechanistic Studies of Molybdenum-Catalyzed Asymmetric Alkylation

David L. Hughes*, Shane W. Krska*, Robert A. Reamer, David J. Mathre, and Yongkui Sun

Department of Process Research, Merck and Company, Inc. Rahway, NJ 07030

Abstract. Structures of key intermediates and delineation of the major features of the catalytic cycle have been determined for the asymmetric Mo-catalyzed asymmetric alkylation. The crystal structure of the π-allyl species was determined, and is characterized by an unusual 3-point binding of an anionic ligand. Based on NMR analysis the structure in solution is consistent with the crystal structure. For the allylic alkylation, the crystal structure predicts the opposite stereochemistry vs. that observed experimentally, which suggests the reaction proceeds via a minor isomer (Curtin-Hammett conditions) or via a retention-retention mechanism. In addition, CO transfer, promoted by $Mo(CO)_6$, has been found to play a key role in catalyst turnover.

Transition-metal catalyzed allylic alkylations have enjoyed widespread use in organic synthesis over the past three decades (1). While palladium has been the metal of choice in the vast majority of these transformations, other metals have found their niche in selected cases. In the 1980's the Trost group (2) discovered that Group VI metals (Mo, W) provided regiochemistry that was complementary to Pd in allylic alkylations, with Mo and W affording the branched product vs. the linear product derived from Pd catalysis (Eq 1).

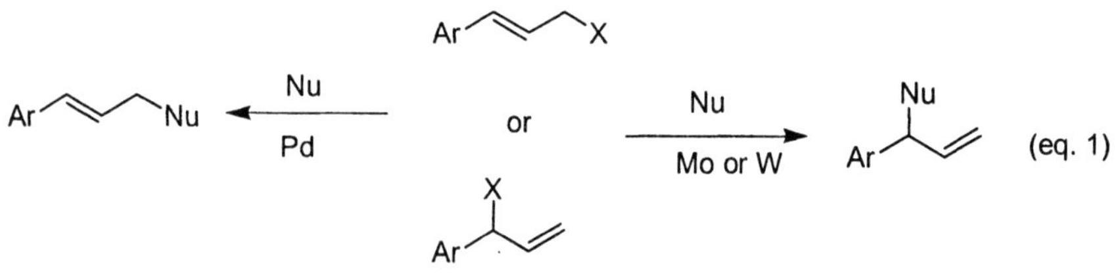

Ar Nu ← Nu / Pd — Ar X or X Ar — Nu / Mo or W → Nu Ar (eq. 1)

Having a chiral center present, the branched products derived from Mo- and W-catalyzed alkylations have the potential for asymmetric induction. In fact, stoichiometric asymmetric reactions with chiral Mo-allyl complexes were realized in the 1980's by Faller (3). The first catalytic asymmetric variant of

Scheme 1

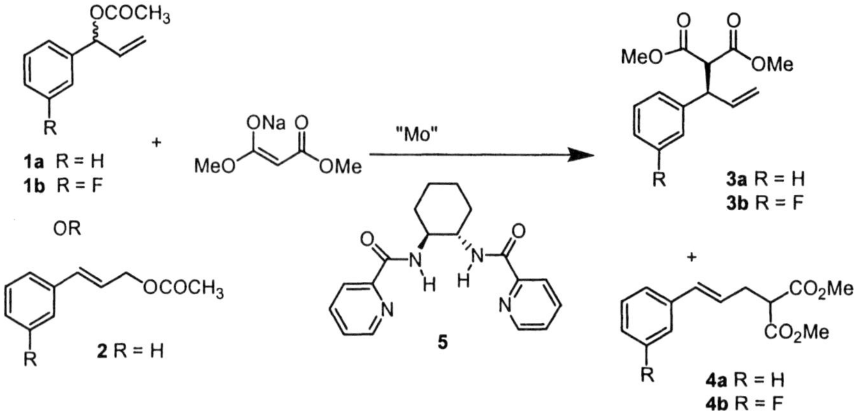

OCOCH$_3$ R 1a R = H 1b R = F OR OCOCH$_3$ R 2 R = H + ONa O MeO OMe "Mo" → O N H H N O N N 5

O O MeO OMe R 3a R = H 3b R = F + CO$_2$Me CO$_2$Me R 4a R = H 4b R = F

the reaction was reported by Pfaltz and Lloyd-Jones in 1995 using a chiral P,N-bidentate ligand and tungsten as the metal catalyst (4). Enantioselectivities of >90% ee were obtained, but the branched/linear ratio was only 3:1. A practical version was subsequently reported by Trost and Hachiya in 1998 using the C_2-symmetrical bis-picolinamide ligand **5** (Scheme 1) (5). This ligand provided high ee's with a variety of aryl substituted allylic carbonates with branched/linear ratios generally >20. A variety of ligands and substrates for this reaction have been recently reported from the laboratories of Kocovsky (6), Pfaltz (7), and Moberg (8).

Our interest in this reaction arose from a need to find an efficient synthesis of the cyclopentanone **6** as a key intermediate in a Merck development candidate. We found the key asymmetric center could be incorporated using the Mo-catalyzed asymmetric alkylation with **1b** as substrate, and the bis-picolinamide **5** as the ligand (5). The synthesis of **6** has been published (9), so further discussion will not be provided here.

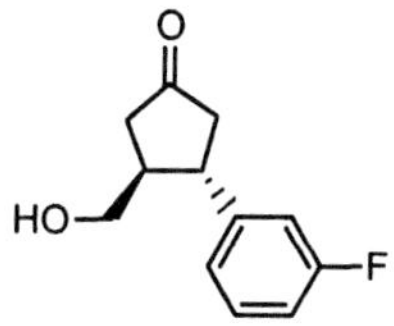

6

One of the main concerns for scaling the Mo-catalyzed alkylation was the molybdenum precatalyst employed. Generally, $Mo(CO)_3$(cycloheptatriene) or $Mo(CO)_3(EtCN)_3$ were used by the Trost, Kocovsky, and Pfaltz groups. These pre-catalysts are not available in quantity, are very expensive, and are quite air-sensitive. The ideal pre-catalyst would be $Mo(CO)_6$, which is cheap, readily available, and generally stable to air and water. At the time we began our work, there were no reports in the literature on the use of $Mo(CO)_6$ in the asymmetric allylic alkylation reaction. Subsequently, Lahred and Moberg (8a) demonstrated $Mo(CO)_6$ could be used in these reactions using microwave irradiation. We found that $Mo(CO)_6$ could be employed if activated thermally. The activation required heating for 4 h in toluene at 90°C or for a similar time in THF at reflux (66°C). When activated in this manner, $Mo(CO)_6$ gave comparable yields, ee's, and branched/linear ratios for a number of substrates relative to the other two Mo precatalysts. This led us to conclude that all three precatalysts were generating the same catalytic species (10).

But what was the nature of this catalytic species? As described in the Trost 1998 communication (5a), the ligand may be binding to the metal in a variety of ways, three of which are depicted below as **A** – **C**.

In **A** ligation occurs in a bidentate fashion, and would involve either one amide group and one pyridine, or the two amide groups. Precedent for this type of coordination is found in the recent work of Lloyd-Jones (11) and Ahn (12), who found structurally similar ligands bound palladium via the amide carbonyl and phosphine (**7a** and **7b**). In **B** ligation involves all four of the nitrogens, and has precedent from the work of Moberg and co-workers, (13) who postulated such binding of the diphenyl-based ligand with zirconium (**8**). In **C**, binding would involve the two pyridine nitrogens ligated to the metal in a trans fashion. When we began our work, no precedent for this sort of binding

A 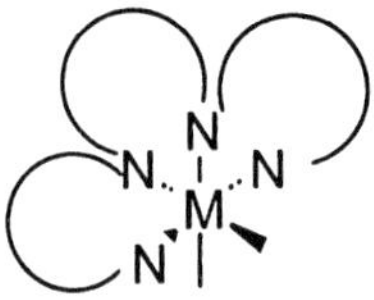 B 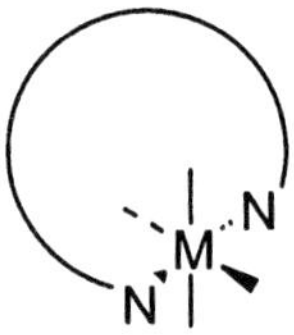 C

mode existed. Since then, potential indirect evidence for this type of coordination has been provided by the work of Kocovsky and coworkers (6a), who found that replacement of a pyridine with phenyl (**9a** to **9b**) resulted in a

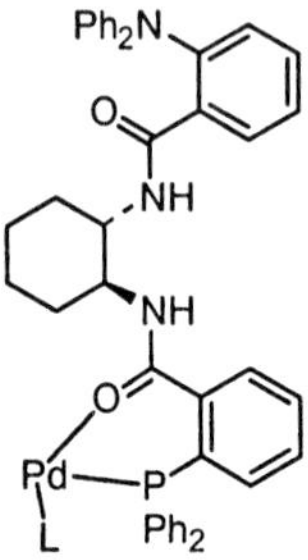

7a 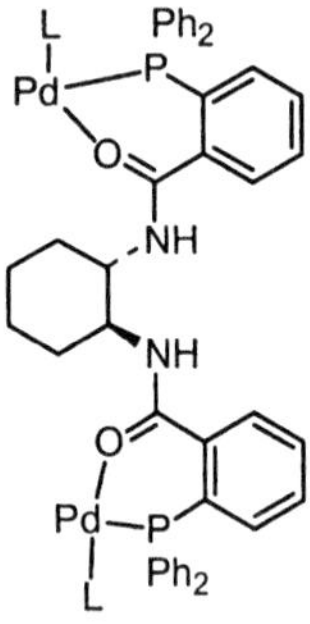 **7b** 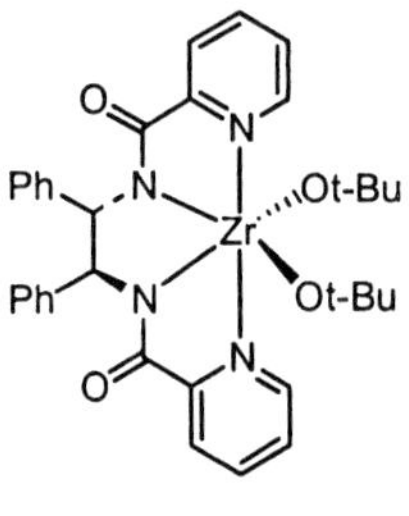 **8** 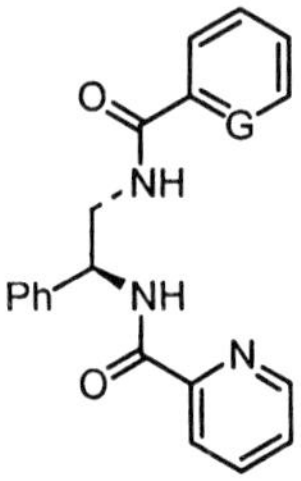 **9a** G = N **9b** G = CH

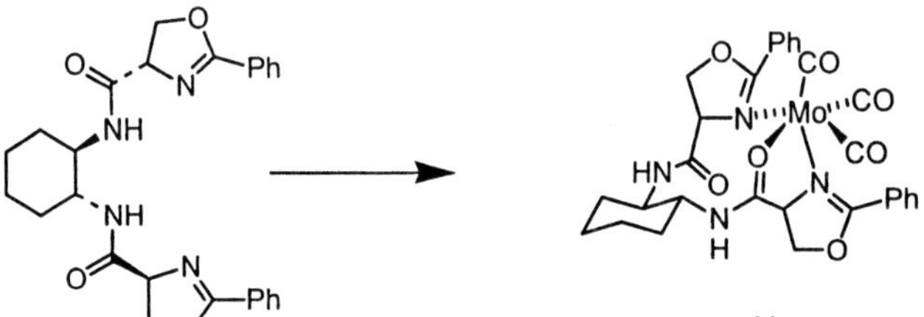

10 **11**

far less active catalyst, although the ee only diminished somewhat. More to the point, in late 2001 Pfaltz published a crystal structure of a Mo(0) complex with the bis(dihydrooxazole) ligand (**10**) with binding via the remote dihydrooxazole nitrogens to form an 11-membered ring (**11**), although the remote nitrogens are bound in a cis-fashion not in the trans-fashion denoted by structure **C** (7b).

Ligand Binding Studied by Spectroscopy and Designed Ligand Probes

Initial NMR Studies

To determine the binding mode involved in the molybdenum-catalyzed reactions, we began with an NMR study using one equiv. each of ligand **5** and (norbornadiene)$Mo(CO)_4$) (hereafter abbreviated as (NBD)$Mo(CO)_4$)in THF-d_8. The equilibrium shown in eq. 2 was established within 10 minutes at ambient

5 —THF, $Mo(CO)_4$Norbornadiene→

^{15}N NMR amide δ 116 ppm
pyridine δ 304 ppm

(Eq. 2)

12 ⇌ **13**

^{15}N NMR amide δ 118, 131 ppm
pyridine δ 304, 274 ppm

^{15}N NMR amide δ 129 ppm
pyridine δ 273 ppm

temperature. Two species were formed, one unsymmetrical (**12**) and one symmetrical (**13**). Diagnostic were the NH's and 6-proton of the pyridine. For the symmetrical species, only one signal was observed for each of the NH's and 6-pyridine protons, and both had moved downfield relative to the uncomplexed ligand. In the unsymmetrical complex, only one NH and one 6-proton moved downfield, while the other NH and 6-proton remained close to the values observed in the neutral ligand. As shown in eq. 2, ^{15}N NMR was also diagnostic, clearly showing that the symmetrical complex involved binding of all 4 nitrogens, while the unsymmetrical complex involved binding of only "half" of the ligand. Addition of more molybdenum precatalyst drove the equilibrium toward the symmetrical product, confirming that it contained two Mo atoms per ligand. Thus, ligand **5** binds Mo via mode **A** and not in the fashion in which the Pfaltz ligand binds (ie, via both remote nitrogens of the heterocycle). In eq. 2, the amide binding is shown via the oxygen, in accordance with the Pd mode of binding in **7a** and **7b**, although by NMR it cannot be established if the binding is through the oxygen or nitrogen.

Designed Ligand Probes

These NMR experiments prompted us to prepare a series of ligands, **14-16**, wherein one or both of the pyridines was replaced with a non-coordinating group, to determine if the binding mode suggested by the NMR experiments translated into actual catalytic activity.

These ligands were studied in the standard allylic alkylation reaction shown in Scheme 1 above (14). The results are tabulated in Table 1. The bis-phenyl ligand **16** had very low activity and produced an ee of only 24%, indicating that binding only via the two amides, if occurring, provided a poor asymmetric catalyst. On the other hand, the picolinamide-benzamide ligand **14** gave results slightly better those of the bis-picolinamide ligand **5**, suggesting binding via one amide and one pyridine was sufficient to generate an active catalyst that affords product with high enantioselectivity. At the same catalyst loading, ligand **14** was about half as reactive as ligand **5**.

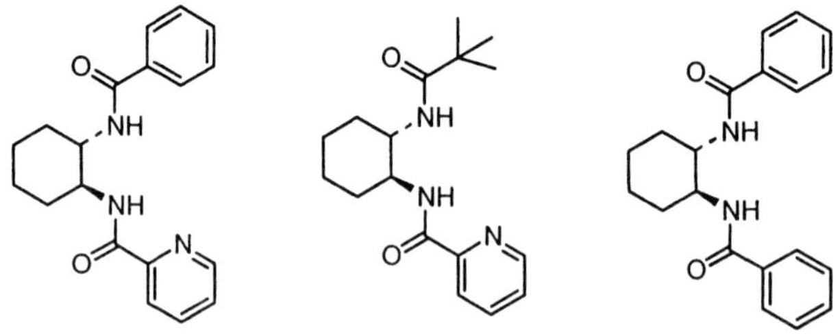

O NH NH O N

14

O NH NH O N

15

O NH NH O

16

Having established that binding via modes **B** and **C** probably does not occur with these ligands, the possibility remains that tri-dentate coordination is occurring involving both amides and the pyridine. To address this possibility, the pivalamide-picolinamide ligand **15** was synthesized with the expectation that if the second amide is involved, increased steric hindrance may diminish its effectiveness as a catalyst. As shown in Table 1, this ligand still gave excellent ee and regioselectivity, although it was about 4-fold less reactive than ligand **5**. This suggests that if the second amide is involved, steric effects are not interfering with formation of a catalytic species capable of producing product with high ee and regioselectivity.

Table 1. Mo-Catalyzed Allylic Alkylation with Designed Ligand Probes

Entry	Ligand	Substrate	Yield	Rel Rate	3/4	% ee
1	**5**	**1a**	86		25	87
2	**5**	**2**	88	1.0	35	97
3	**14**	**1a**	90		53	92
4	**14**	**2**	90	0.5	60	99
5	**15**	**1a**	88		28	88
6	**15**	**2**	95	0.25	30	98
7	**16**	**1a**	35	<0.02	1	24

Conditions. $(C_7H_8)Mo(CO)_3$ precatalyst; sodium dimethyl malonate was prepared *in situ* from dimethyl malonate and NaH, combined with the allylic carbonate, and added to the activated catalyst solution in THF at 48°C

Additional NMR experiments on ligand binding

Given the complexity of binding with ligand **5**, where both 1:1 and 2:1 Mo:ligand complexes are formed (eq. 2), all further NMR studies were carried out with ligand **14**, where only the 1:1 Mo:ligand binding mode is possible.

The complexation of ligand **14** with three Mo precatalysts was studied to determine if the precatalysts were in fact producing the same catalytic complex, as our synthetic studies previously suggested (10). As shown in Scheme 2, the $Mo(CO)_6$ and $(NBD)Mo(CO)_4$ precatalysts produce the same complex **(18)** with 4 CO's bound to Mo and two-point binding to the ligand. The main difference is that refluxing in THF is required to displace the two CO's from $Mo(CO)_6$, while the norbornadiene precursor forms complex **18** within a few minutes at room temperature since displacement of CO is not required.. For (cycloheptatriene)$Mo(CO)_3$, the ligand binds in a similar two-point fashion, but an equilibrium exists between the complex with 3 and 4 CO's (**17** /**18** ratio of 4/1). The other ligand in complex **17** is not defined, but may be a THF molecule. These results confirm the reaction results reported earlier (10) that concluded all Mo precatalysts make the same catalytically active species.

Formation of the π-allyl intermediate

With the mode of binding of the ligands with molybdenum established, we next focused on the subsequent step in the reaction sequence, the oxidative addition of the Mo(0) species to the allylic carbonate to generate the π-allyl intermediate. Reaction of the neutral Mo-ligand complex **18** with linear carbonate **2** in THF-d_8 at 50 °C in a sealed NMR tube resulted in formation of the π-allyl intermediate **19**. The stoichiometry of this reaction, as shown in eq. 3, involves formation of one equivalent each of **19**, MeOH and free ligand **14** from 2 equivalents of complex **18**. In addition, a significant amount (presumably

Scheme 2

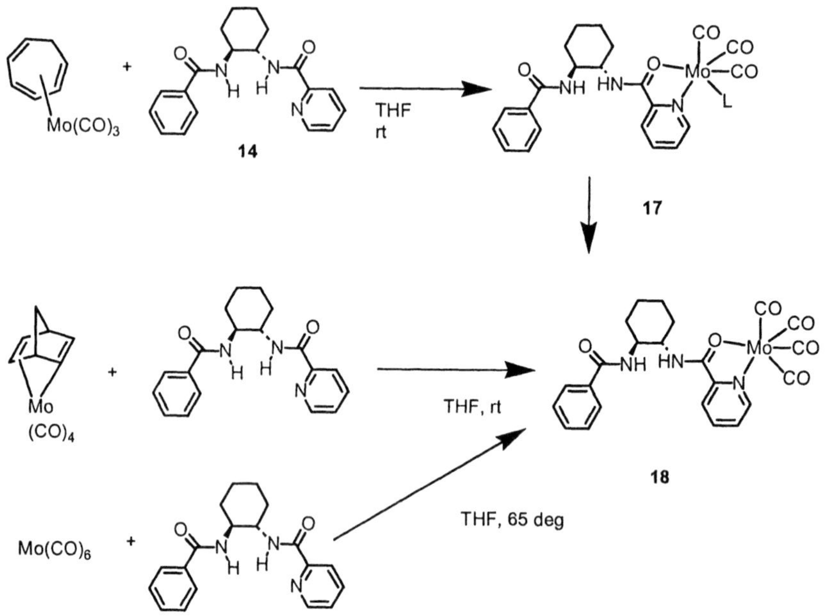

$Mo(CO)_3$ + 14 → THF rt → 17 (Mo with CO, CO, CO, L)

17 → 18

$(CO)_4$Mo (norbornadiene) + 14 → THF, rt → 18 (Mo with CO, CO, CO, CO)

$Mo(CO)_6$ + 14 → THF, 65 deg → 18

1 equivalent) of $Mo(CO)_6$ is observed by ^{13}C NMR spectroscopy. The methanol observed is formed, along with CO_2, from the methyl carbonate leaving group upon deprotonation of one of the amide groups of the ligand; the resulting anionic ligand ultimately becomes a part of the π-allyl complex **19**.

A suitable crystal of the π-allyl complex was grown and the X-ray structure is shown in Figure 1. Noteworthy features of this structure include the following: (1) the allyl moeity binds in a η^3 fashion to Mo, with one face clearly open for reaction with a nucleophile; (2) the ligand is coordinated to the metal via three-point binding: the pyridine nitrogen, the nitrogen of the deprotonated amide, and the carbonyl oxygen of the undeprotonated amide; and (3) the complex contains two CO ligands. The overall geometry of the complex, including the *syn* orientation of the allyl moiety with respect to the CO ligands, is similar to other structurally characterized $L_2(CO)_2XMo(\eta^3\text{-allyl})$ complexes (15).

2 **18** $\xrightarrow[\textbf{2}]{\text{THF}}$ **19** + $Mo(CO)_6$ + MeOH + CO_2 + free ligand **14** (Eq. 3)

The solution structure, as determined by multinuclear NMR and depicted in Figure 2, is consistent with the crystal structure. Key features of the solution structure include the following: (1) The allyl moiety is in the W-conformation, as confirmed by nOe's between H^a and H^d, and H^b and H^c. (2) H^b and H^c show nOe's to the 6-proton of the pyridine ring, indicating they face the pyridine group. (3) H^a and H^d correlate to both CO's, consistent with these protons being on the opposite side and facing the CO's. (4) In the ^{15}N NMR spectrum, the amide nitrogen chemical shifts are at δ 130 and 175 ppm and the pyridine nitrogen at δ 263 ppm in comparison to the free ligand where the respective shifts are δ 118, 119 and 304 ppm. The 175 ppm resonance for one of the amide nitrogens, the absence of one NH proton in the ^1H spectrum, as well as proton coupling data, clearly indicate one amide is deprotonated in the complex. The other amide is still protonated, but the downfield shift of the nitrogen in the ^{15}N NMR and the NH proton in ^1H NMR vs. the uncomplexed ligand suggests this amide is also involved in coordination to Mo in the solution complex, as is also observed in the crystal structure.

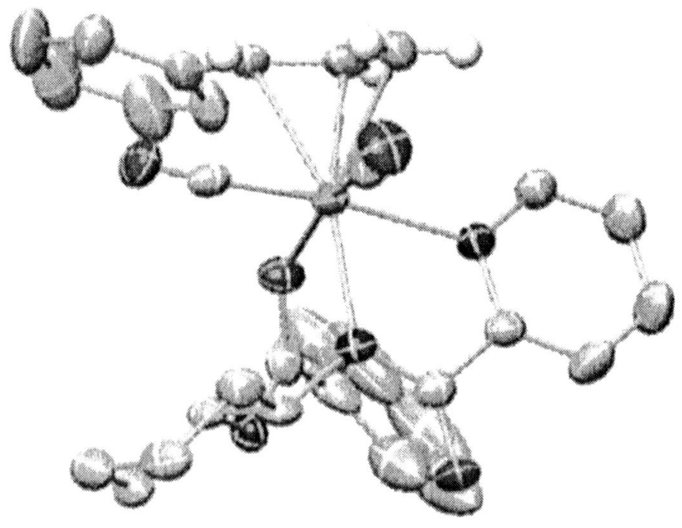

Figure 1. ORTEP plot of the crystal structure of complex **9**. Color coding: Molybdenum is green, carbon gray, hydrogen white, nitrogen blue, and oxygen red. (See page 2 of color insert.)

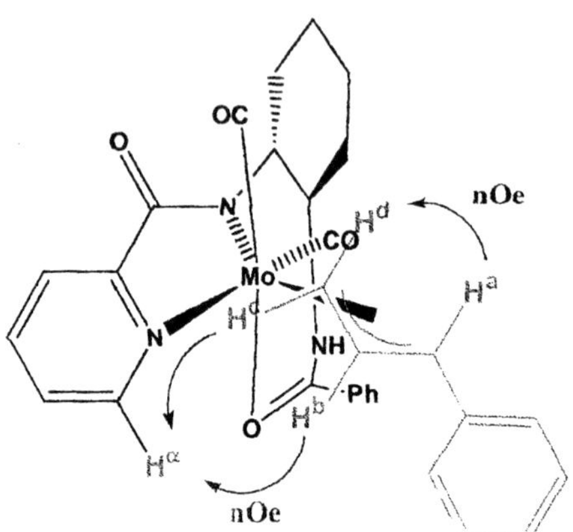

Figure 2. Solution structure of the π-allyl complex **19** in THF-d_8 (See page 2 of color insert.)

Two important and unexpected findings of the structure are the deprotonation of the amide nitrogen and the three-point binding of the ligand to Mo. Both are likely key factors in providing strong complexation and the rigidity required for producing high ee's in the allylic alkylation, even at temperatures of >100 °C in toluene (16). While the crystal structure indicates one face of the π-allyl moiety is clearly open for approach of a nucleophile, attack of malonate in this direction predicts stereochemistry opposite to that observed. Thus, the π-allyl species observed as the major species in solution may not be the reactive species (Curtin-Hammett conditions) or the displacement with malonate may be occurring via an intramolecular (retention) process. The most recent results (17) from our laboratories suggest that the overall reaction is occurring via a retention-retention process.

Reactivity of π-Allyl Intermediate 19

Having prepared and characterized the π-allyl intermediate **19**, we next investigated its reactivity with sodium dimethyl malonate. When the π-allyl intermediate was generated in solution from complex **18** and the allylic carbonate **2**, it reacted cleanly with sodium dimethyl malonate ($NaCHE_2$) to form the substituted malonate product with an ee of >95%. However, when the isolated crystals of complex **19** were redissolved in THF and combined with $NaCHE_2$, surprisingly, no reaction occured. The difference between the two modes of preparation is clear from the stoichiometry shown in eq. 3. The solution

preparation of complex **19** also includes formation of an additional mole of free ligand, MeOH, and $Mo(CO)_6$. We felt it likely that $Mo(CO)_6$ was playing a key role in the reactivity, and this was confirmed when addition of $Mo(CO)_6$ to the unreactive solution of complex **19** and $NaCHE_2$ resulted in initiation of the alkylation. The role of the $Mo(CO)_6$ became clear upon identifying the molybdenum-containing product (**20**) of the alkylation reaction, which is shown in eq. 4. This product was identified by NMR and also isolated and fully characterized. It is simply the N-deprotonated analog of complex **18**. The role $Mo(CO)_6$ plays in this reaction is as a carbonyl shuttle. The π-allyl intermediate **19** contains 2 CO's, while the product of the reaction, **20**, contains 4 CO's. For the reaction to initiate, a CO source is required to activate complex **19**, and that is accomplished by $Mo(CO)_6$. We have also found that running the reaction under 1 atm of CO also initiates the reaction, although the rate is slower than with $Mo(CO)_6$.

Completion of the Catalytic Cycle

Complex **20** can be prepared by reaction of complex **18** with NaH or $NaCHE_2$. For isolation purposes, it was prepared from NaH. Isolated **20** was

Scheme 3. Catalytic Cycle

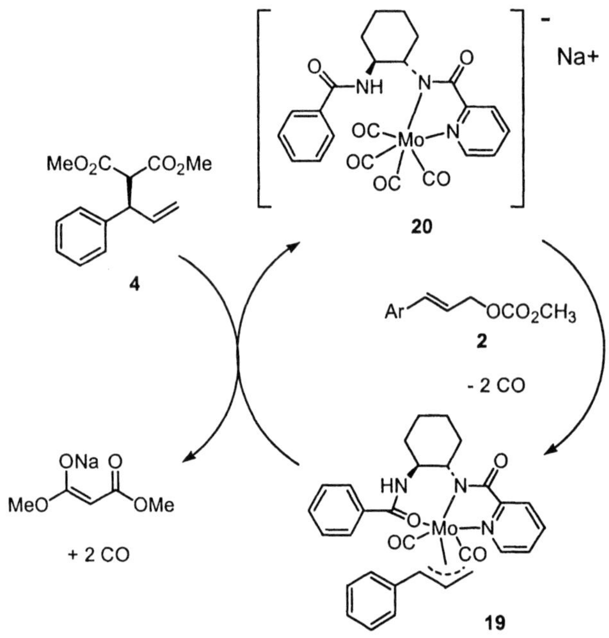

dissolved in THF and reacted with carbonate **2** to generate the π-allyl species **19**. This completes the catalytic cycle, as shown in Scheme 3, which involves the following steps. Precatalyst **18** reacts with carbonate **2** to generate π-allyl complex **19** according to eq. 3. Complex **19** reacts with malonate in the presence of a CO source (either **18** or $Mo(CO)_6$) to form product **4** and molybdate complex **20**. Complex **20** reacts with carbonate **2** to regenerate complex **19** and release CO. A typical synthetic experiment using catalytic molybdenum was followed by NMR, and complexes **19** and **20** were the only Mo-containing species observed, with high mass balance in Mo conserved throughout the reaction. This confirms that **19** and **20** are the catalyst resting states under actual catalytic synthetic conditions.

Summary

The mechanism of the Mo-catalyzed allylic alkylation has been delineated via isolation and characterization of the key intermediates in the catalytic cycle. In the neutral pre-complex of ligand **5** and Mo, coordination occurs via one amide and one pyridine of the ligand. This led to design of the picolinamide-benzamide ligand **14**, which gave excellent yields, ee's, and branched/linear ratios in the allylic alkylation. The π-allyl intermediate **19** derived from ligand **14** and the methyl carbonate of cinnamyl alcohol was crystallized. The structure of the π-allyl species is characterized by an unusual 3-point binding of an anionic ligand. Based on NMR analysis the structure in solution is consistent with the crystal structure. In addition, CO transfer, promoted by $Mo(CO)_6$, has been found to play a key role in catalyst turnover.

References

1. Trost, B. M.; van Vranken, D. L.; *Chem. Rev.* 1996, *96*, 395. Hayashi, T. In *Catalytic Asymmetric Synthesis*; Ojima, I. Ed.; VCH Publishers: New York, 1993; pp. 325-365. Consigio, G.; Waymouth, R. M. *Chem. Rev.* 1989, *89*, 257.
2. Molybdenum: (a) Trost, B. M.; Lautens, M. *J. Am. Chem. Soc.* **1982**, *104*, 5543; (b) *ibid*, **1983**, *105*, 3344; (c) *ibid*, **1987**, *109*, 1469; (d) *Tetrahedron* **1987**, *43*, 4817. (e) Trost, B. M; Merlic, C. A. *J. Am. Chem. Soc.* **1990**, *112*, 9590. Tungsten: Trost, B. M.; Hung, M.-H. *J. Am. Chem. Soc.* **1983**, *105*, 7757; **1984**, *106*, 6837. Trost, B. M.; Tometzki, G. B.; Hung, M.- H. *J. Am. Chem. Soc.* **1987**, *109*, 2176.

3. Faller, J. W.; Mazzieri, M. R.; Nguyen, J. T.; Parr, J.; Tokunaga, M. *Pure Appl. Chem.* **1994**, *66*, 1463. Faller, J. W.; Chao, K.-H. *J. Am. Chem. Soc.* **1983**, *105*, 3893. Faller, J. W.; Lambert, C.; Mazzieri, M. R. *J. Organomet. Chem.* **1990**, *383*, 161.
4. Lloyd-Jones, G. C.; Pfaltz, A. *Angew. Chem. Int. Ed. Engl.* **1995**, *34*, 462-464.
5. Trost, B. M.; Hachiya, I. *J. Am. Chem. Soc.* **1998**, *120*, 1104-1105. Trost, B. M.; Hildbrand, S.; Dogra, K. J. Am. Chem. Soc. **1999**, *121*, 10416-10417.
6. Malkov, A. V.; Spoor, P.; Vinader, V.; Kočovský, P. *Tetrahedron. Lett.* **2001**, *42*, 509-512. Kocovsky, P.; Malkov, A. V.; Vyskocil, S.; Lloyd-Jones, G. C. *Pure Appl. Chem.* **1999**, *71*, 1425-1433.
7. Glorius, F.; Pfaltz, A. *Org. Lett.* **1999**, *1*, 141-144; Glorius, F.; Neuburger, M.; Pfaltz, A. *Helv. Chim. Acta* **2001**, *84*, 3178-3196.
8. Kaiser, N.-F. K.; Bremberg, U.; Larhed, M.; Moberg, C.; Hallberg, A. *Angew. Chem. Int. Ed. Engl.* **2000**, *39*, 3596-3598. Belda, O.; Kaiser, N.-F.; Bremberg, U; Larhed, M.; Hallberg, A.; Moberg, C. *J. Org. Chem.* **2000**, *65*, 5868-5870.
9. Palucki, M.; Um, J. M.; Yasuda, N.; Conlon, D. A.; Tsay, F.-R.; Hartner, F. W.; Hsiao, Y.; Marcune, B.; Karady, S.; Hughes, D. L.; Dormer, P. G.; Reider, P. J. J. Org. Chem. 2002, 67, in press.
10. Palucki, M.; Um, J. M.; Conlon, D. A.; Yasuda, N.; Hughes, D. L.; Mao, B.; Wang, J.; Reider, P. J. *Adv. Synth. Catal.* **2001**, *343*, 46-50.
11. Butts, C. P.; Crosby, J.; Lloyd-Jones, G. C.; Stephen, S. C. *Chem. Commun.* **1999**, 1707-1708.
12. Kim, Y. K.; Lee, S. J.; Ahn, K. H. *J. Org. Chem.* **2000**, *65*, 7807-7813.
13. Adolfsson, H.; Moberg, C. *Tetrahedron: Asymm.* **1995**, *6*, 2023-2031.
14. Trost, B. M.; Dogra, K.; Hachiya, I.; Emura, T.; Hughes, D. L.; Krska, S.; Reamer, R. A.; Palucki, M. P.; Yasuda, N.; Reider, P. J. *Angew. Chem. Int. Ed. Engl.* **2002**, *41*, 1929-1932.
15. Curtis, M. D.; Eisenstein, O. Organometallics **1984**, 3, 887-895.
16. Hughes, D. L.; Palucki, M.; Yasuda, N.; Reamer, R. A.; Reider, P. J. *J. Org. Chem.* **2002**, *67*, 2762-2768.
17. Lloyd-Jones, G. C.; Krska, S. W.; Hughes, D. L.; Gouriou, L.; Bonnet, V. D.; Jack, K.; Sun, Y.; Reamer, R. A. *J. Am. Chem. Soc*, **2004**, *126*, in press.

Chapter 10

Asymmetric Catalysis in Ionic Liquids: Easy Recycling of Catalyst and Improvement of Catalytic Performances

Choong Eui Song

Life Sciences Division, Korea Institute of Science and Technology, P.O. Box 131, Cheongryang, Seoul 130–650, Korea

Ambient temperature ionic liquids have recently come to be regarded as an eco-friendly alternative to volatile organic solvents in chemical processes, due to their negligible vapor pressure and high thermal stability. In these solvents, chiral catalysts having polar or ionic character can be immobilized without additional structural modification and thus the ionic solutions containing the catalyst can easily be separated from the reagents and reaction products, and then, be reused. More interestingly, switching from an organic solvent to an ionic liquid often results in an improvement in catalytic performance (e.g., rate acceleration, enantioselectivity improvement and an increase in catalyst stability).

During the last decade, a number of powerful catalytic asymmetric reactions have emerged as a result of the growing need to develop more efficient and practical synthetic methods for biologically active compounds (1). Although a number of homogeneous chiral catalysts have gained wide acceptance in terms of their efficiency and selectivity, the contribution of asymmetric catalysis in the overall production of chiral chemicals is much lower than originally expected.

The high cost and toxicity of these catalysts and the possible contamination of the product with the catalyst has restricted the use of asymmetric catalytic reactions in industry. Therefore, the development of an efficient immobilization method for chiral homogeneous catalysts is highly desirable, since immobilized catalysts offer several practical advantages over soluble catalysts, such as the easier separation of the expensive and/or toxic catalyst from the reagents and reaction products, the simplification of the methods for catalyst recycling, and the possible adaptation of the immobilized catalyst to continuous-flow processes. For catalyst recycling, homogeneous chiral catalysts can be immobilized either by anchoring the catalyst on a solid support or by using an aqueous or fluorous biphasic system (2). All of these approaches are interesting, but usually require additional modifications to the catalyst structure. Moreover, such approaches frequently lead to partial loss of activity and/or enantioselectivity. Recently, a new approach has been adopted for catalyst separation and recycling involving the use of ionic liquids, (3) i.e., salt mixtures with a melting point below ambient temperature. Air and moisture stable room temperature ionic liquids consisting of 1,3-dialkylimidazolium cations and their counter anions, in particular, have attracted much interest over the last few years. In these solvents, catalysts having a polar or ionic character can be immobilized without additional structural modification and thus the ionic solutions containing the catalyst can easily be separated from the reagents and reaction products, and then, be reused.

Considering the practical importance of catalyst recovery, and the fact that ionic liquids are expected to become an alternative medium for immobilization, surprisingly little attention has been devoted to performing asymmetric catalytic reactions in ionic liquids. Over the last few years, therefore, we have become interested in the immobilization of chiral catalysts using ionic liquids, and found that ionic liquids can act as powerful media in many asymmetric catalytic reactions, not only for facilitating catalyst recovery, but also for improving catalytic performance (rate acceleration, selectivity improvement and increased catalyst stability, etc.) (4).

In this article, our recent work on the use of ionic liquids for the immobilization of homogeneous chiral catalysts is presented. The ionic liquids used in these studies are mainly 1-butyl-3-methylimidazolium salts **1** (5).

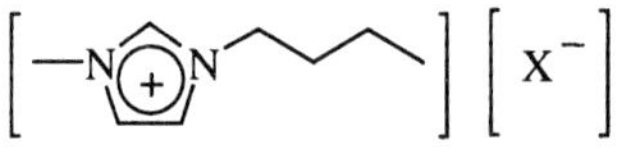

[bmim][X] (**1**)

X = PF_6 (**1a**), SbF_6 (**1b**), BF_4 (**1c**), OTf (**1d**), $N(Tf)_2$ (**1e**)

Os-Catalyzed Asymmetric Dihydroxylation of Olefins in Ionic Liquids

The Sharpless Os-catalyzed asymmetric dihydroxylation (AD) of olefins provides one of the most elegant methods for the synthesis of chiral vicinal diols (6). Although the AD reaction offers a number of processes that could be applied to the synthesis of chiral drugs, natural products, fine chemicals, etc., the high cost and toxicity of osmium, and possible contamination of the product with osmium catalyst in the product have restricted the industrial use of the AD reaction. In order to explore the possibility of both catalytic components being used repetitively, several attempts to immobilize this catalytic system have been made (7-10). Early approaches to immobilizing OsO_4 on solid-supported alkaloid ligands suffered from several disadvantages, such as the need of complicated synthesis of the supported ligand system and a reduction of catalytic efficiencies. Moreover, the effective recovery of osmium has failed in all cases, since the coordination of the anchored ligands and the osmium tetroxide is in equilibrium. On the other hand, the alternative immobilization of the osmium catalyst by microencapsulation of OsO_4 in a polymer matrix (8), by using an ion-exchange technique (9) or by osmylation of macroporous resins bearing residual vinyl groups such as Amberlite XAD-4 (10), provides a recoverable and reusable system for the osmium catalyzed AD reaction.

Our quite recent work (11) demonstrates that the combination of the ionic liquid [bmim][PF_6] **1a** or [bmim][SbF_6] **1b** and a new bis-cinchona alkaloid **2** generated in situ from $(QN)_2$PHAL [1,4-bis(9-*O*-quininyl)phthalazine] (12) during the AD reaction (Scheme 1) provides a simple and highly practical approach to the immobilization of both catalytic components (osmium and the alkaloid ligand), based on the immobilization of the osmium-ligand catalyst in the ionic liquid phase. Initially, to investigate the effect of an ionic liquid on the AD reaction, as well as the recyclability of the catalytic components, the AD reactions of olefins were carried out with the well-known ligand, 1,4-bis(9-*O*-dihydroquininyl)-phthalazine [$(DHQ)_2$PHAL], using the standard Upjohn conditions (13) (*N*-methylmorpholine-*N*-oxide (NMO) as a cooxidant) in the presence of the [bmim][PF_6] ionic liquid **1a** at 20 °C. The results obtained in the presence of the ionic liquid **1a** were quite comparable to those obtained without an ionic liquid. For example, the AD reaction of *trans*-stilbene was completed within 20 min, affording the corresponding diol in 95% yield with 85% ee (the olefin was added in one portion). Encouraged by this result, we next performed a catalyst recycling experiment as follows: After completion of the reaction, all the volatiles were removed under reduced pressure and the diol produced was extracted from the residue with pre-cooled (0 °C) diethyl ether. The remaining ionic liquid phase was then subjected to the next run with a new batch of olefin and NMO, without the addition of any osmium or ligand. However, further reuse of the recovered ionic liquid phase resulted in a dramatic decrease in the yield

(45% after 24 h) of the product, due to severe leaching of both osmium and $(DHQ)_2PHAL$ during the extraction with ether. Leaching of the catalytic components during the extraction can be ascribed to the partial solubility of $(DHQ)_2PHAL$ in ether. Since the complex formation of OsO_4 and an alkaloid ligand is expected to be reversible, the lowering of the concentration of the chiral ligand in the ionic liquid phase might result in more leaching of OsO_4 from the ionic liquid phase. Therefore, we assumed that the use of an alkaloid ligand, which can be strongly immobilized in an ionic liquid, can minimize Os leaching during the extraction of the product. To prove our assumption, we used 1,4-bis(9-*O*-quininyl)phthalazine [$(QN)_2PHAL$] as the ligand. $(QN)_2PHAL$ will be converted to the alkaloid **2** bearing highly polar residues (four hydroxy groups) during the AD reactions of olefins (Scheme 1).

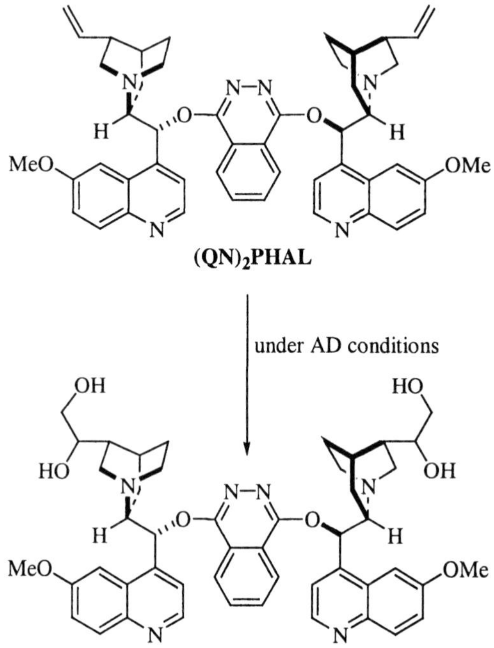

$(QN)_2PHAL$

under AD conditions

2

Scheme 1

The use of $(QN)_2PHAL$ instead of $(DHQ)_2PHAL$ afforded the same yields and ees (e.g., 95% yield, 97% ee for *trans*-stilbene; 96% yield, 94% ee for methyl *trans*-cinnamate; reaction conditions: 1 mol% of OsO_4, 5 mol% of $(QN)_2PHAL$, 2 equiv of the [bmim][PF_6] **1a** and slow addition of olefin for 12 h) and, moreover, resulted in a drastic improvement in the recyclability of both catalytic components. The recovered ionic liquid phase containing both osmium and the ligand **2** could be recycled several times, even in the recycle experiments using 0.1 mol% of OsO_4 (Scheme 2). In the case of the recycle experiments using 0.1 mol% of OsO_4, the total turnover number (TON) was 2370. To the best of our knowledge, this is the highest TON value ever reported under Upjohn conditions. Thus, our procedure can allow the minimization of catalyst consumption and accordingly, the reduction of the osmium contamination both in the product and the work-up waste. Moreover, the $(QN)_2PHAL$ ligand is much more economic to prepare than conventional AD ligands, such as $(DHQ)_2PHAL$ (12).

Contemporaneously, Afonso et al. (14) also developed a recyclable catalyst system for the AD reaction using the ionic liquid **1a**. They found that, in the [bmim][PF_6] ionic liquid, AD reactions using $K_3Fe(CN)_6$ as the cooxidant can also be carried out and that, after the reaction, the ionic solution containing both osmium and ligand can be recovered and reused several times.

OsO_4
$(QN)_2PHAL$ (5 mol%)
NMO(1.5 equiv)
[bmim][PF_6]
acetone-H_2O (10:1), 20 ¡É

OH
OH

Using 1 mol% of OsO_4:
92%, 98% ee (1st run); 88%, 96% ee (2nd run); 91%, 94% ee (3rd run)
70%, 94% ee (4th run); 50%, 94% ee (5th run)
Using 0.1 mol% of OsO_4:
90%, 98% ee (1st run); 89%, 92% ee (2nd run); 58%, 89% ee (3rd run)

Scheme 2

Chiral (Salen)Mn(III) Complex Catalyzed Asymmetric Epoxidation of Olefins in an Ionic Liquid

Catalytic asymmetric epoxidation of alkenes presents a powerful strategy for the synthesis of enantiomerically enriched epoxides. Amongst several possible catalytic methods, the asymmetric epoxidation of unfunctionalized alkenes catalyzed by chiral Mn(III)(salen) complexes such as **3**, developed by Jacobsen and his coworkers, is one of the most reliable methods (15). In particular, very high enantioselectivities are obtained for *cis*-disubstituted and *tri*-substituted olefins.

Although many attempts have been made to immobilize Jacobsen's catalyst, including the covalent attachment of the complex to insoluble supports (16), the ion exchange of Mn(III) complexes into the intra-crystalline space of zeolites or mesoporous materials (17), the steric occlusion in the nano-sized cages of zeolites using a "ship in a bottle methodology" (18), physical entrapment in a polydimethylsiloxane membrane (19), and the utilization of a fluorous biphasic system (FBS) (20), etc., the results that have been obtained up until now are far from satisfactory.

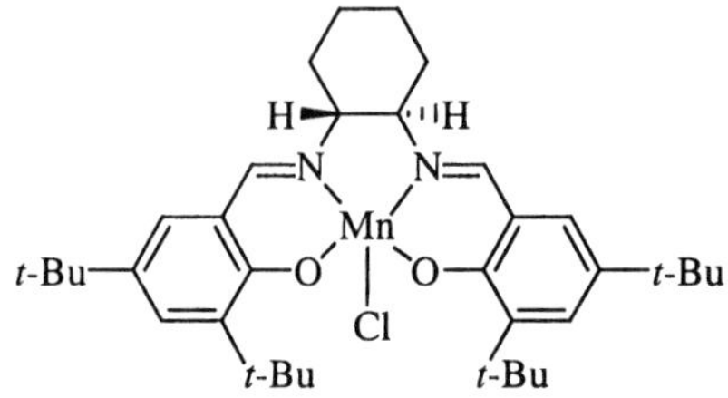

(*R*,*R*)-**3**

Our recently published work (21) indicates that the use of an ionic liquid is also of advantage in this important reaction. We investigated the epoxidation of various olefins (2,2-dimethylchromene, 6-cyano-2,2-dimethylchromene, indene, *cis*-β-methylstyrene, and 1-phenylcyclohexene) in the presence of a chiral Mn(III)(salen) catalyst (*R*,*R*)-**3** using NaOCl as cooxidant in a mixture of [bmim][PF_6] **1a** and CH_2Cl_2 (1:4 v/v). Good conversion and enantioselectivity were observed for all tested substrates (e.g., 86% yield and 96% ee for 2,2-dimethylchromene). More interestingly, we obtained a clear enhancement of catalytic activity by adding the ionic liquid to the organic solvent. The epoxidation of 2,2-dimethylchromene using 4 mol% of catalyst in the presence of the ionic liquid **1a** was completed in 2 h. However, the same reaction without **1a** required 6 h to achieve complete conversion. This rate acceleration effect of the ionic liquid was shown more dramatically when the amount of catalyst was reduced to 0.5 mol% (Figure 1) (22). In a mixture of [bmim][PF_6] **1a** and CH_2Cl_2 (1:4 v/v), the reaction was completed in 6 h, whereas without the ionic liquid only ca. 40% of conversion was observed after the same reaction time. Moreover, the use of the ionic liquid solvent allows for easy catalyst recycling, without the need for any catalyst modification. By washing the organic phase with water followed by extraction of the product with hexane, the ionic catalyst solution is recovered after reaction and can be reused (Scheme 3). However, the enantioselectivity and activity of the recovered catalyst decreased slightly after sequential use. After being recycled five times, the yield and enantioselectivity dropped from 83 % to 53 % and from 96% ee to 88% ee, respectively, under identical reaction conditions. This might be due to a minor degradation of the salen catalyst (*R*,*R*)-**3** under oxidation conditions. Nevertheless, to the best of our knowledge, Jacobsen's catalyst immobilized in an ionic liquid constitutes one of the most efficient and recyclable catalytic systems available for the asymmetric epoxidation of alkenes.

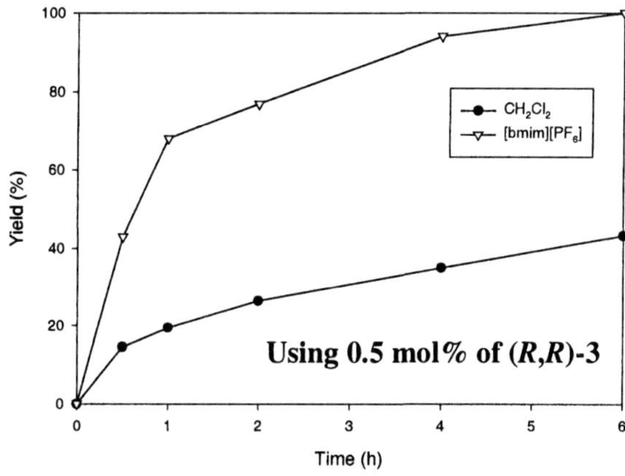

Figure 1. Kinetic studies in the epoxidation of 2,2'-dimethylchromene in the presence of 0.5 mol% of (*R*,*R*)-**3** using NaOCl as cooxidant in CH_2Cl_2 or [bmim][PF_6]/CH_2Cl_2 (1:4 v/v) at 0 °C.

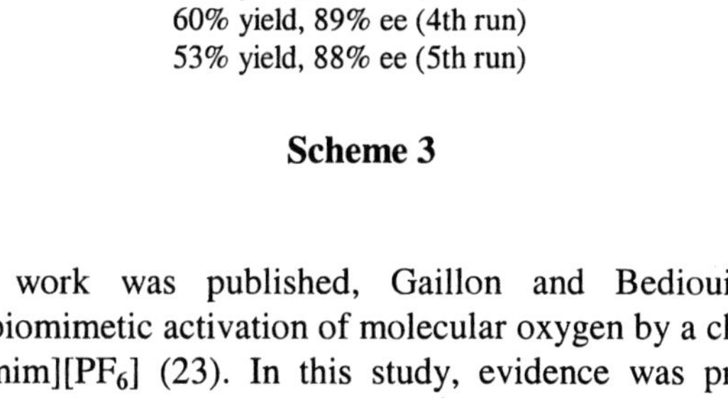

(*R*,*R*)-**3** (4 mol%)
NaOCl
[bmim]PF_6-CH_2Cl_2
(1 : 4, v/v), 0 °C, 2 h

86% yield, 96% ee (1st run)
73% yield, 90% ee (2nd run)
73% yield, 90% ee (3rd run)
60% yield, 89% ee (4th run)
53% yield, 88% ee (5th run)

Scheme 3

After our work was published, Gaillon and Bedioui reported the electroassisted biomimetic activation of molecular oxygen by a chiral Mn (salen) complex in [bmim][PF_6] (23). In this study, evidence was provided for the formation of the highly reactive $[Mn(V)=O]^+$ manganese-oxo intermediate that could transfer its oxygen to an olefin. This method might have the potential to be used for electrocatalytic asymmetric epoxidation with molecular oxygen in ionic liquid media. However, no experimental data on a preparative scale was reported.

Chiral Cr(Salen) Complex Catalyzed Asymmetric Ring Opening of Meso-Epoxides with Azide in Ionic Liquids

Asymmetric ring opening reactions (ARO) of epoxides with trimethylsilyl azide ($TMSN_3$) catalyzed by the chiral Cr(salen) complex **4** has been recognized as an attractive approach to the synthesis of optically pure β-amino alcohols (24). Especially, the chiral Cr(salen) catalyst **4** exhibited indefinite stability under catalytic conditions which allowed for its repeated recycling. Jacobsen reported that this reaction can be run without solvents and the catalyst can be recycled a number of times without loss of activity and enantioselectivity (25). However, this catalyst recycling procedure involves the potentially hazardous distillation of the neat liquid azides which can not apply to large scale applications.

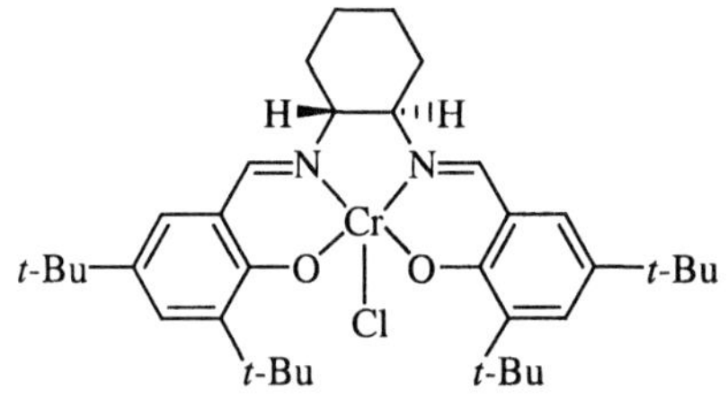

(*R*,*R*)-**4**

We recently developed a highly practical recycling procedure for the chiral Cr(salen) catalyst **4** involving the use of [bmim] salts **1a-d** (26). Our procedure consists of running the reaction of $TMSN_3$ with *meso*-epoxides in the presence of catalytic amounts of (*R*,*R*)-**4** dissolved in the [bmim] ionic liquids. As shown in Table 1, the yield and enantioselectivity are strongly dependent upon the nature of the counteranion: while the reaction performed in hydrophobic [PF_6] **1a** and [SbF_6] **1b** salts gave high yields and degrees of enantioselectivity (similar to those obtained in organic solvents), the system is almost inactive when performed in hydrophilic [BF_4] **1c** and [OTf] **1d** salts. Although, using the hydrophobic ionic liquids, excellent results were achieved, the catalyst existed as a suspended form in the ionic liquids, when hexane was added to the reaction mixture after reaction. On the other hand, although the reaction hardly occurred in the hydrophilic ionic liquids, the catalyst was formed a clear red-brown solution phase, which can make its separation from the hexane phase more easily. Thus, the best recyclable catalytic system was obtained by immobilizing the catalyst in a mixture (5/1 v/v) of hydrophobic [bmim][PF_6] and hydrophilic [bmim][OTf] ionic liquids. The recovered ionic liquid phase containing the catalyst was reused several times without any loss of activity and enantioselectivity even after the fifth use (Scheme 4). This recycling procedure does not have any hazardous work-up procedure, such as the distillation of the azide product, and moreover, provides the additional advantage of being able to use a catalyst without any modification of the structure.

Table 1. Chiral Cr(salen) Complex Catalyzed Asymmetric Ring Opening of Meso- Epoxides with $TMSN_3$ in Ionic Liquids

TMSN$_3$, (*R*,*R*)-**4** (3 mol%), ionic liquid **1**, 20 °C → N_3 / OTMS

Substrate (Y)	Ionic Liquid	t/h	Yield (%)	% ee
CH_2	**1a**	28	76	94
CH_2	**1b**	28	75	87
CH_2	**1c**	28	5	3
CH_2	**1d**	28	trace	-
$(CH_2)_2$	**1a**	18	86	85
O	**1a**	18	74	97

(*R*,*R*)-**4** (3 mol%), $TMSN_3$

[bmim][PF_6]/[bmim][OTf] (5:1, v/v)

20 °C

N_3 OTMS

68% yield, 94% ee (1st run)
72% yield, 93% ee (2nd run)
85% yield, 93% ee (3rd run)
75% yield, 94% ee (4th run)
76% yield, 93% ee (5th run)

Scheme 4

Jacobsen's Chiral Co(III)(salen) Complex Catalyzed Hydrolytic Kinetic Resolution of Racemic Epoxides in the Ionic Liquid

Hydrolytic kinetic resolution (HKR) of racemic epoxides using Jacobsen's chiral (salen)Co(III)(OAc) complex **5**·OAc as a catalyst is one of the most practical approaches to the preparation of enantiopure terminal epoxides (24,27). The chiral catalyst **5**·OAc is readily accessible, and displays high enantioselectivity. However, this catalyst provides relatively low turnover numbers and frequencies. To facilitate catalyst separation and catalyst reuse, some attempts to anchor Jacobsen's catalyst onto insoluble supports have already been made (28). Although these heterogeneous analogues of **5**·OAc gave almost the same enantioselectivities as compared to those of the homogeneous one, complicated synthetic manipulations were required for their preparation. Moreover, during the reaction, these solid-bound catalysts (28), as well as the homogeneous ones (27a) are reduced to Co(II) complex **5** which is known to be inactive for HKR (27a) and thus, they need to be re-oxidized to Co(III) complex with acetic acid under air before being used in the next run.

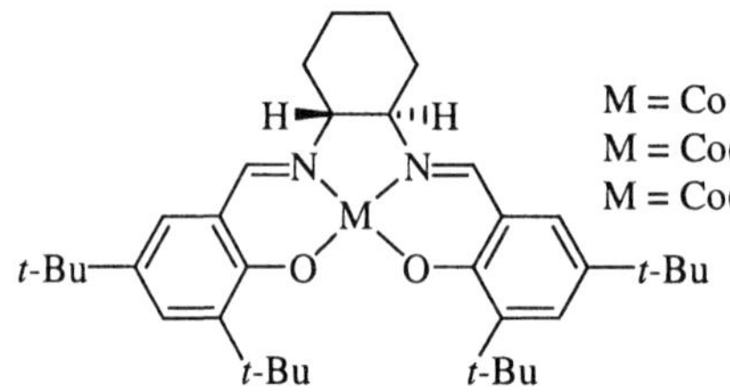

H H N N M O O t-Bu t-Bu t-Bu t-Bu

M = Co: (*R*,*R*)-**5**
M = Co(OAc): (*R*,*R*)-**5**-OAc
M = Co(PF_6): (*R*,*R*)-**5**-PF_6

Quite recently, we came upon a very interesting effect produced by ionic liquids on this catalytic reaction (29). In the chiral Co(III)(salen)-catalyzed HKR of racemic epoxides using catalytic amounts of (*R*,*R*)-Co(III)(salen) complex (*R*,*R*)- **5**·OAc in a mixture (4/1, v/v) of THF and an ionic liquid, [bmim][X] **1** (X = PF_6 (**1a**), NTf_2 (**1e**)), at 20 °C, the yields and enantiomeric excesses were also quite comparable to those (27a) obtained without ionic liquids (Scheme 5)

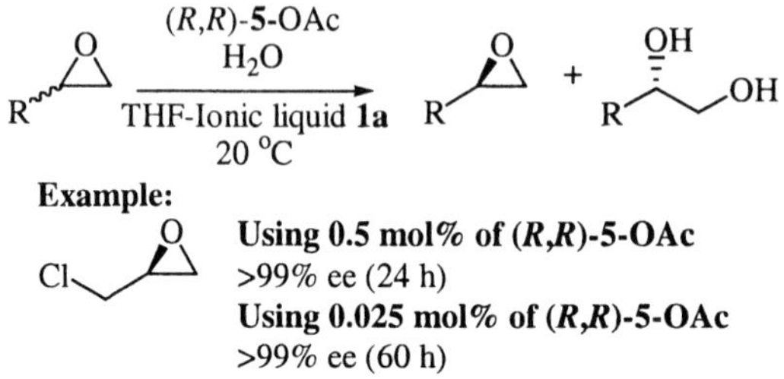

(*R*,*R*)-**5**-OAc
H_2O
THF-Ionic liquid **1a**
20 °C

Example:

Using 0.5 mol% of (*R*,*R*)-5-OAc
>99% ee (24 h)
Using 0.025 mol% of (*R*,*R*)-5-OAc
>99% ee (60 h)

Scheme 5

More importantly, we found that the oxidation state of the Co(salen) complex dissolved in the recovered ionic liquid phase was not +II, but +III. As mentioned above, in the case of using organic solvents as the reaction media, the Co(III) catalyst **5**·OAc is reduced to Co(II)-complex **5** during the HKR reactions. The UV spectrum of the ionic liquid phase containing Co(salen) complex (recovered after the HKR of epichlorohydrin in [bmim][PF_6] (**1a**)/THF) is the same as that of Co(III)(salen)(PF_6) complex **5**·PF_6 (30). On the other hand, the UV spectrum of the Co complex recovered after the HKR carried out only in THF solvent is similar to that of Co(II) complex **5** (Figure 2-1). In the XPS (X-ray Photoelectron Spectroscopy) spectrum of both the recovered Co(salen) complex and Co(III)(salen)(PF_6) complex **5**·PF_6, the Co $2p_{3/2}$ line appeared at ~780 eV, whereas the XPS line of Co(II)(salen) complex **5** appeared at ~777 eV (Figure 2-2). These results clearly indicate that the recovered cobalt complex has a +III oxidation state.

Moreover, we also found that in the presence of the ionic liquid **1a** or **1e**, catalytically inactive Co(II)(salen) complex **5** instead of Jacobsen's chiral Co(III)(salen)·OAc catalyst **5**·OAc can be directly used as a catalyst precursor. The Co(II) complex **5** is oxidized, without the use of acetic acid, to catalytically active Co(III) complex during the HKR reactions, which may not be possible in conventional organic solvents. All HKRs of racemic epichlorohydrin using catalytic amounts of Co(II) complex (*R*,*R*)-**5** proceeded smoothly, even using 0.025 mol% of (*R*,*R*)-**5**. For example, enantiomerically pure epichlorohydrin was obtained after 70 h using 0.025 mol% of catalyst. Moreover, the catalytically active Co(III) oxidation state is stabilized against reduction to the Co(II) complex, which enables the reuse of the recovered catalyst for subsequent runs without additional reoxidation. This catalytic system involving the ionic liquid [bmim][NTf_2] **1e** was able to be reused up to ten times without any loss of activity and enantioselectivity (>99% ee) (Scheme 6). Very interestingly, the catalytic activity of the recovered ionic liquid phase increased upon reuse (reaction time; 22 h for the first run and 2 h for the tenth run). The reason for the increase in activity upon reuse can be ascribed to the increasing concentration of catalytically active Co(III) complex in the reaction mixture. However, at the present time, it is not clear why, in the presence of an ionic liquid, Co(II) complex was oxidized, without the use of acetic acid, to catalytically active Co(III) complex and this oxidation state was maintained. We are currently carrying out research in an attempt to understand the above mentioned effects of ionic liquids on the Co(salen)-catalyzed HKR reactions of racemic epoxides.

(*R*,*R*)-**5** (0.5 mol%)
H_2O
THF-[bmim][$N(Tf)_2$]
20 °C
>99% ee even after
10th use of catalyst

Scheme 6

The catalyst stabilization effect of the ionic liquid was also observed in the asymmetric hydrogenation of olefins. Guernik et al. (31) recently reported that Rh-MeDuPHOS complex immobilized in [bmim][PF_6] (**1a**) catalyzes the asymmetric hydrogenation of enamides (methyl α-acetamidoacrylate and methyl α-acetamidocinnamate) with enantioselectivities similar to those obtained using

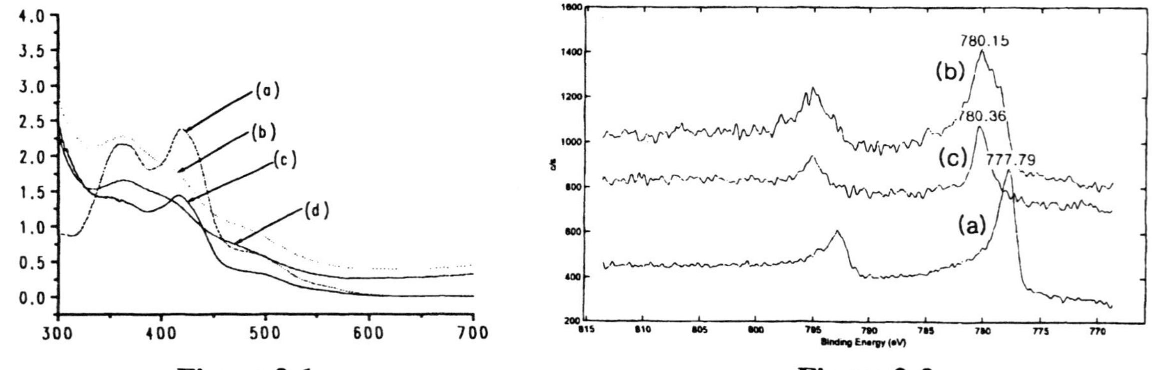

Figure 2-1

Figure 2-2

Figure 2. UV and XPS Spectra of various Co(salen) complexes: **Figure 2-1**; (a) the Co(II) complex (*R*,*R*)-**5**, (b) the Co(III)(PF_6) complex (*R*,*R*)-**5**-PF_6, (c) the recovered Co(salen) complex after HKR of epichlorohydrin in THF, (d) the ionic liquid phase containing Co(salen) complex recovered after HKR of epichlorohydrin in [bmim][PF_6]/THF. **Figure 2-2**; (a) Co(II)(salen) complex (*R*,*R*)-**5**, (b) Co(III)(salen) complex (*R*,*R*)-**5**-PF_6, (c) the recovered Co(salen) complex after HKR of epichlorohydrin in [bmim][PF_6]/THF.

the same catalyst dissolved in organic solvent (*i*-PrOH). Interestingly, the ionic liquid provided extra stability for this highly air-sensitive catalyst, so that all experiments involving catalyst recycling could be carried out under an air atmosphere without any significant loss of enantioselectivity (methyl α-acetamidocinnamate as the substrate: 96% ee for the first run and 94% ee for the fifth run) (Scheme 7). On the other hand, in the absence of the ionic liquid, the catalyst prepared in an inert atmosphere and then exposed to air for a few minutes showed almost no catalytic activity. The stabilization effect of the ionic liquid was considered to be due to the entrapment of the air-sensitive complex in an ionic liquid.

Finally, it should be mentioned here that a significant increase of enantioselectivity in the ionic liquid **1a** was also observed in the enantioselective Pd-catalyzed allylic substitution. Pd-ferrocenylphosphine complexes, such as BPPFA-Pd complex, catalyzed the allylic substitution of (*rac*)-(*E*)-1,3-diphenyl-3-acetoxypro-1-ene in **1a** afforded the product with 68 % ee. This ee value is much higher than that (40% ee) obtained in THF (Scheme 8) (32).

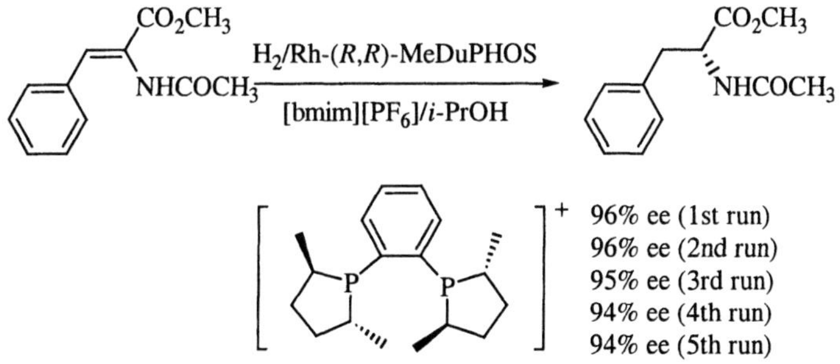

CO_2CH_3 / $NHCOCH_3$ — H_2/Rh-(*R*,*R*)-MeDuPHOS, [bmim][PF_6]/*i*-PrOH → CO_2CH_3 / $NHCOCH_3$

96% ee (1st run)
96% ee (2nd run)
95% ee (3rd run)
94% ee (4th run)
94% ee (5th run)

(*R*,*R*)-MeDuPHOS

"All manipulations were performed in air."

Scheme 7

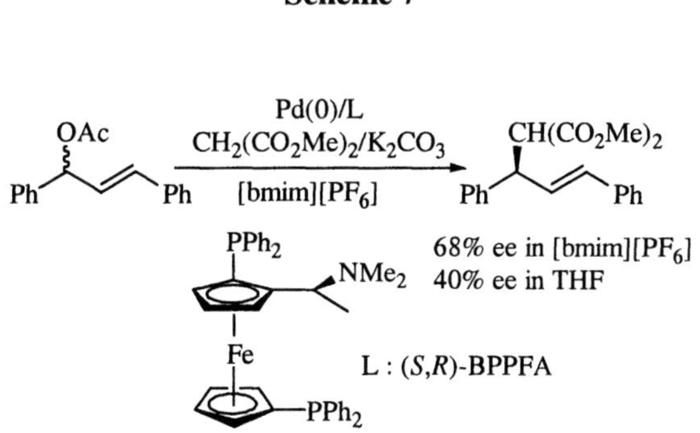

OAc, Ph, Ph — Pd(0)/L, $CH_2(CO_2Me)_2/K_2CO_3$, [bmim][PF_6] → $CH(CO_2Me)_2$, Ph, Ph

PPh_2, NMe_2, Fe, PPh_2

68% ee in [bmim][PF_6]
40% ee in THF

L : (*S*,*R*)-BPPFA

Scheme 8

Conclusion

In conclusion, the examples presented here show that the use of ionic liquids offers many advantages in asymmetric catalysis. The use of the ionic liquid solvent allows the catalyst to be recycled more easily, without the need for any catalyst modification. Moreover, in some cases, the ionic liquids were observed to have a positive influence on the catalytic properties. The reactions can be accelerated in suitable ionic solutions and can proceed with improved selectivities. The increased stability of the catalyst is also observed in the ionic liquid. In some cases, the activity of the catalyst increased rather upon subsequent reuse. However, most reactions are strongly influenced by the structure of the ionic liquids, especially by the nature of the counter anion. Accordingly, choosing the most suitable ionic liquid for a specific reaction is very important. A more detailed understanding of these interesting phenomena needs to be obtained in the future. Nevertheless, I believe that immobilization methodology involving the use of ionic liquids can open up new perspectives for the immobilization of chiral catalysts.

Acknowledgements

The financial support provided by the Ministry of Science and Technology in Korea (NRL-program) and the Korea Institute of Science and Technology is gratefully acknowledged.

References

1. (a) *Comprehensive Asymmetric Catalysis*; Jacobsen, E. N.; Pfaltz, A.; Yamamoto, H., Eds.; Springer: Berlin, 1999; Vol. 1-3. (b) *Catalytic Asymmetric Synthesis*; Ojima, I., Ed.; VCH: New York, 2000.
2. (a) *Chiral Catalyst Immobilization and Recycling*; de Vos, D. E.; Vankelecom, I. F. J.; Jacobs, P. A., Eds.; Wiley-VCH: New York, 2000. (b) Song, C. E.; Lee, S.-g. *Chem. Rev.* **2002**, *102*, 3495. (c) Fan, Q.-H.; Li, Y.-M.; Chan, A. C. S. *Chem. Rev.* **2002**, *102*, 3385.
3. For recent reviews on ionic liquids, (a) Zhao, H.; Malhotra, S. V. *Aldrichimica Acta* **2002**, *35*, 75. (b) Dupont, J.; de Souza, R. F.; Suarez, P. A. Z. *Chem. Rev.* **2002**, *102*, 3667. (c) Sheldon, R. *Chem. Commun.* **2001**, 2399. (d) Wasserscheid, P.; Keim, W. *Angew. Chem. Int. Ed. Engl.* **2000**, *39*, 3772. (e) Welton, T. *Chem. Rev.* **1999**, *99*, 2071.
4. We have recently observed the effects of ionic liquids on the rate acceleration and selectivity enhancement also in some non-asymmetric catalytic reactions, e.g., Friedel-Crafts alkylation of arenes,[4a] Diels-Alder reaction,[4b] and nucleophilic substitutions,[4c,d] etc. (a) Song, C. E.; Shim. W.

H.; Roh, E. J.; Choi, J. H. *Chem. Commun.* **2000**, 1695. (b) Song, C. E.; Shim, W. H.; Roh, E. J.; Lee, S.-g.; Choi, J. H. *Chem. Commun.* **2001**, 1122. (c) Kim, D. W.; Song, C. E.; Chi, D. Y. *J. Am. Chem. Soc.* **2002**, *124*, 10278. (d) Kim, D. W.; Song, C. E.; Chi, D. Y. *J. Org. Chem.* **2003**, *68*, in press.

5. Ionic liquids used in our study were nearly chloride-free (<5 ppm) and purchased from C-TRI Co., Ltd. (www.c-tri.com).
6. For reviews: (a) Kolb, H. C.; Van Nieuwenhze, M. S.; Sharpless, K. B. *Chem. Rev.* **1994**, *94*, 2483. (b) Markó, I. E.; Svendsen, J. in *Comprehensive Asymmetric Catalysis II*; Jacobsen, E. N.; Pfaltz, A.; Yamamoto, H. Eds.; Springer Verlag: Berlin, **1999**; Vol. 2, pp 713.
7. (a) Salvadori, P.; Pini, D.; Petri, A. *Synlett* **1999**, 1181. (b) Bolm, C.; Gerlach, A. *Eur. J. Org. Chem.* **1998**, *21*, 21.
8. (a) Nagayama, S.; Endo, M.; Kobayashi, S. *J. Org. Chem.* **1998**, *63*, 6094. (b) Kobayashi, S.; Endo, M.; Nagayama, S. *J. Am. Chem. Soc.* **1999**, *121*, 11229. (c) Kobayashi, S.; Ishida, T.; Akiyama, R. *Org. Lett.* **2001**, *3*, 2649.
9. (a) Choudary, B. M.; Chowdari, N. S.; Kantam, M. L.; Raghavan, K. V. *J. Am. Chem. Soc.* **2001**, *123*, 9220. (b) Choudary, B. M.; Chowdari, N. S.; Jyothi, K.; Kantam, M. L. *J. Am. Chem. Soc.* **2002**, *124*, 5341.
10. Yang, J. W.; Han, H.; Roh, E. J.; Lee, S.-g.; Song, C. E. *Org. Lett.* **2002**, *4*, 4685.
11. Song, C. E.; Jung, D.; Roh, E. J.; Lee, S.-g.; Chi, D. Y. *Chem. Commun.* **2002**, 3038.
12. Song, C. E.; Yang, J. W.; Ha, H. J.; Lee, S.-g. *Tetrahedron: Asymmetry* **1996**, *7*, 645.
13. Van Rheenen, V.; Kelly, R. C.; Cha, P. Y. *Tetrahedron Lett.* **1976**, *23*, 1973.
14. Branco, L. C.; Afonso, C. A. M. *Chem. Commun.* **2002**, 3036.
15. Jacobsen, E. N.; Wu, M. H. in *Comprehensive Asymmetric Catalysis*; Jacobsen, E. N.; Pfaltz, A.; Yamamoto, H., Eds.; Springer Verlag: Berlin, 1999; Vol. 2, pp 649.
16. Selected recent papers: (a) Canali, L.; Cowan, E.; Deleuze, H.; Gibson, C. L.; Sherrington, D. C. *Chem. Commun.* **1998**, 2561. (b) Kim, G.-J.; Shin, J.-H. *Tetrahedron Lett.* **1999**, *40*, 6827. (c) Pini, D.; Mandoli, A.; Orlandi, S.; Salvadori, P. *Tetrahedron: Asymmetry* **1999**, *10*, 3883. (d) Song, C. E.; Roh, E. J.; Yu, B. M.; Chi, D. Y.; Kim, S. C.; Lee, K.-J. *Chem. Commun.* **2000**, 615.
17. Selected papers: (a) Piaggio, P.; McMorn, P.; Murphy, D.; Bethell, D.; Page, P. C. B.; Hancock, F. E.; Sly, C.; Kerton, O. J.; Hutchings, G. J. *J. Chem. Soc., Perkin Trans 2* **2000**, 2008. (b) Kim, G.-J.; Kim, S.-H. *Catal. Lett.* **1999**, *57*, 139. (c) Fraile, J. M.; García, J. I.; Massam, J.; Mayoral, J. A. *J. Mol. Catal. A: Chemical* **1998**, *136*, 47.

18. (a) Sabater, M. J.; Corma, A.; Domenech, A.; Fornes, V.; Garcia, H. *Chem. Commun.* **1997**, 1285. (b) Ogunwumi, S. B.; Bein, T. *Chem. Commun.* **1997**, 901.
19. (a) Vankelecom, I. F. J.; Tas, D.; Parton, R. F.; Van de Vyver, V.; Jacobs, P. A. *Angew. Chem. Int. Ed. Engl.* **1996**, *35*, 1346. (b) Janssen, K. B. M.; Laquiere, I.; Dehaen, W.; Parton, R. F.; Vankelecom, I. F. J.; Jacobs, P. A. *Tetrahedron: Asymmetry* **1997**, *8*, 3481.
20. (a) Pozzi, G.; Cinato, F.; Montanari, F.; Quici, S. *Chem. Commun.* **1998**, 877. (b) Cavazzini, M.; Manfredi, A.; Montanari, F.; Quici, S.; Pozzi, G. *Chem. Commun.* **2000**, 2171.
21. Song, C. E.; Roh, E. J. *Chem. Commun.* **2000**, 837.
22. Our unpublished result.
23. Gaillon, L.; Bedioui, F. *Chem. Commun.* **2001**, 1458.
24. (a) Jacobsen, E. N. *Acc. Chem. Res.* **2000**, *33*, 421. (b) Jacobsen, E. N.; Wu, M. H. in *Comprehensive Asymmetric Catalysis*; Jacobsen, E. N.; Pfaltz, A.; Yamamoto, H., Eds.; Springer-Verlag: Berlin, 1999; Vol. 3, pp 1309.
25. Martínez, L. E.; Leighton, J. L.; Carsten, D. H.; Jacobsen, E. N. *J. Am. Chem. Soc.* **1995**, *117*, 5897.
26. Song, C. E.; Roh, E. J. *Chem. Commun.* **2000**, 1743.
27. (a) Tokunaga, M.; Larrow, J. F.; Kakiuchi, F.; Jacobsen, E. N. *Science* **1997**, *277*, 936. (b) Schaus, S. E.; Brandes, B. D.; Larrow, J. F.; Togunaga, M.; Hansen, K. B.; Gould, A. E.; Furrow, M. E.; Jacobsen, E. N. *J. Am. Chem. Soc.* **2002**, *124*, 1307.
28. (a) Annis, D. A.; Jacobsen, E. N. *J. Am. Chem. Soc.* **1999**, *121*, 4147. (b) Kim, G.-J.; Park, D.-W. *Catal. Today* **2000**, *63*, 537.
29. Oh, C. R.; Choo, D. J.; Shim, W. H.; Lee, D. H.; Roh, E. J.; Lee, S.-g.; Song, C. E. *Chem. Commun.* **2003**, 1100.
30. Co(III)(salen)(PF_6) **1**·PF_6 was prepared by reaction of Co(II)(salen) complex **1** with $(C_5H_5)_2Fe(PF_6)$ according to the following Katsuki's procedure: Irie, R.; Noda, K.; Ito, Y.; Matsumoto, N.; Katsuki, T. *Tetrahedron: Asymmetry* **1991**, *2*, 481.
31. Guernik, A.; Wolfson, A.; Herskowitz, M.; Greenspoon, N.; Geresh, S. *Chem. Commun.* **2001**, 2314.
32. Toma, S.; Gotov, B.; Kmentova, I.; Solcaniova, E. *Green Chem.* **2000**, *2*, 149.

Chapter 11

Phosphite Ligands in Asymmetric Hydrogenation

Montserrat Diéguez, Oscar Pàmies, Aurora Ruiz, and Carmen Claver*

Departament de Química Física i Inorgànica, Universitat Rovira i Virgili, Pl. Imperial Tàrraco, 1, 43005 Tarragona, Spain

Phosphites are extremely attractive ligands for transition metal catalysts. Their available synthesis allows the modification of electronic and steric properties providing the synthesis of series of chiral ligands that can lead to a successful ligand optimization. This chapter describe the use of phoshite ligands in the metal-catalyzed asymmetric hydrogenation of C=C and C=N, which are two of the major homogeneous processes in asymmetric synthesis.

The asymmetric hydrogenation of prochiral compounds catalyzed by chiral transition-metal complexes has been widely used in stereoselective organic synthesis and some processes have found industrial applications (1,2,3). For many years, the scope of this reaction has been gradually extended in both reactant structure and catalyst efficiency. A large number of chiral ligands, mainly P- and N-containing ligands with either C2- or C1-symmetry, have been successfully applied (1,2,3). Diphosphines have played a dominant role among the P-ligands, but recently a group of less electron-rich phosphorus compounds —phosphite ligands—has received much attention. Phosphite ligands have been successfully applied in many other transition-metal-catalysed reactions such as hydroformylation (4,5,6,7,8,9), hydrocyanation (10) and allylic alkylation (11,12,13,14). These ligands are extremely attractive for catalysis because they are easy to prepare from readily available alcohols. The availability of many alcohols makes simple ligand tuning possible, allowing the synthesis of many series of chiral ligands that can be screened in the search for high activity and selectivity. Another advantage of phosphite ligands is that they are less sensitive to air and other oxidizing agents than phosphines. On the other hand, phosphites are more prone to side reactions like hydrolysis, alcoholysis, and the Arbuzov reaction. These side reactions are minimized, however, when bulky aryl phosphites are used.

In this chapter, we report the opportunities that these ligands provide for improving the performance of asymmetric hydrogenation.

Ligand Synthesis

In general, phosphite ligands are synthesized very efficiently in two steps from the corresponding alcohols (Scheme 1). The first step is the formation of the corresponding phosphorochloridite, usually by reaction of a diol with PCl_3 and a base. This phosphorochloridite then reacts with a new alcohol in basic media to form the desired monophosphite or diphosphite ligand.

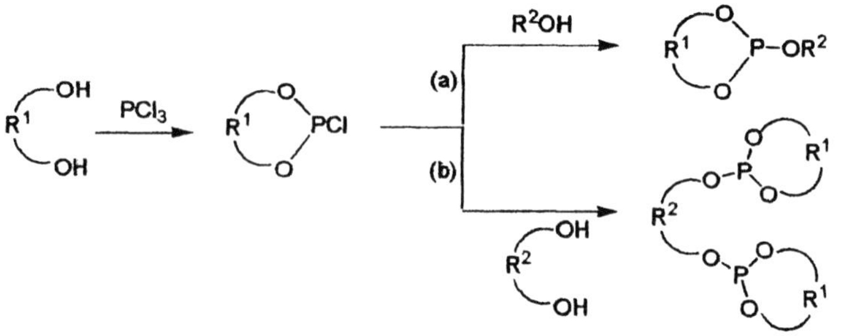

Scheme 1. Typical synthesis of (a) monophosphite and (b) diphosphite ligands.

Hydrogenation of carbon-carbon double bonds

The hydrogenation of carbon-carbon double bonds is widely used for the preparation of high value compounds that can be used as interesting building blocks for asymmetric synthesis. For instance, the hydrogenation of α-dehydroaminoacid derivatives and enamides provides access to unnatural aminoacids and amines that are useful intermediates for the pharmaceutical and agrochemical industries (1-3). The hydrogenation of α-dehydroaminoacid derivatives is also a typical reaction for testing the efficiency of new chiral ligands.

In this section we describe the results published for the asymmetric hdrogenation of C=C bonds with di- and mono-phosphite ligands.

Diphosphite ligands

The first reports on the use of diphosphite ligands in asymmetric hydrogenation were introduced by Brunner (15), Wink (16) and Kolich (17). Brunner and Wink used diphosphite ligands based on carbohydrate and tartaric acid derivatives in the Rh-catalyzed hydrogenation of enamides and obtained low enantioselectivies (1-34 % ee) (Figure 1). Kolich used chiral diphosphites based on optically active binapthol in the Ru-catalyzed hydrogenation of 2-(4-isobutyl-phenyl)acrylic acid and also obtained low ee's (7-34 %) (figure 1).

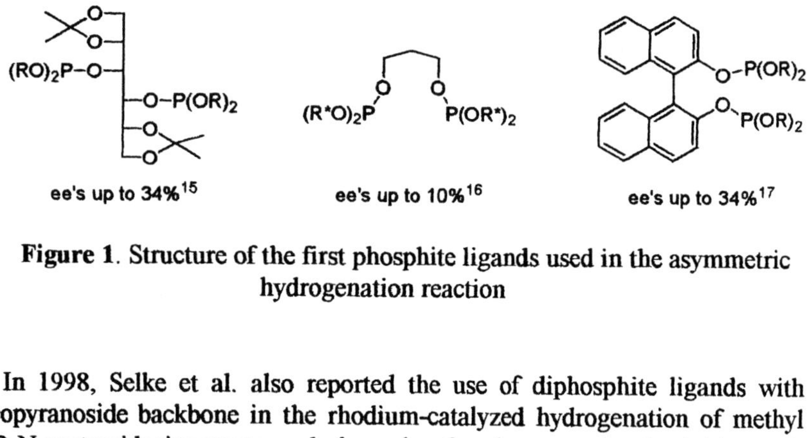

(RO)$_2$P–O– –O–P(OR)$_2$ (R*O)$_2$P P(OR*)$_2$ O–P(OR)$_2$ O–P(OR)$_2$

ee's up to 34%[15] ee's up to 10%[16] ee's up to 34%[17]

Figure 1. Structure of the first phosphite ligands used in the asymmetric hydrogenation reaction

In 1998, Selke et al. also reported the use of diphosphite ligands with glucopyranoside backbone in the rhodium-catalyzed hydrogenation of methyl (*Z*)-2-N-acetamidocinnamate and showed rather low enantioselectivities (ee's up to 13%) (Figure 2) (18).

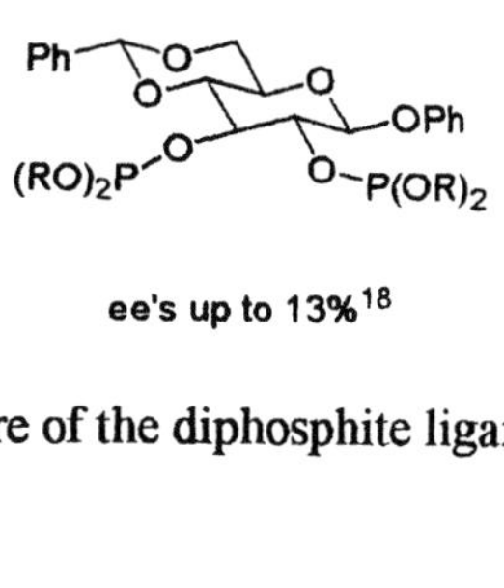

Ph OPh (RO)$_2$P O–P(OR)$_2$

ee's up to 13%[18]

Figure 2. Structure of the diphosphite ligands used by Selke

An important breakthrough in the use of phosphite ligands for asymmetric hydrogenation came with the work of Reetz and coworkers (19). These authors developed a series of C2 derivative ligands derived from mannitol with different phosphite substituents (**a-e**) (Figure 3).

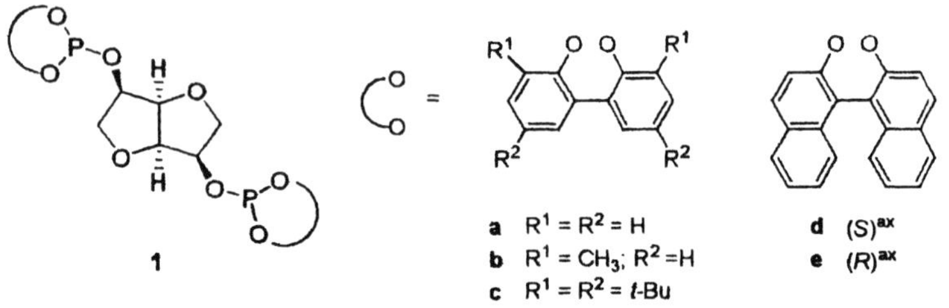
1
a R¹ = R² = H
b R¹ = CH₃; R² =H
c R¹ = R² = t-Bu
d (S)ax
e (R)ax

Figure 3. D-Mannite diphosphite ligands developed by Reetz et al.

These ligands were efficiently applied in the Rh-catalyzed hydrogenation of dimethyl itaconate (Table 1) and methyl *N*-acetamidoacrylate.

Table 1. Rh-catalyzed hydrogenation of dimethyl itaconate using phophite ligands 1.[a]

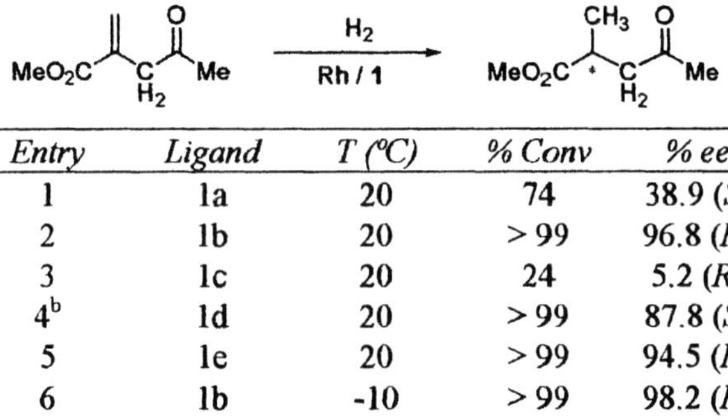
MeO₂C
H₂
Me
H₂
Rh / 1
CH₃
MeO₂C
H₂
Me

Entry	*Ligand*	*T (°C)*	*% Conv*	*% ee*
1	1a	20	74	38.9 (*S*)
2	1b	20	>99	96.8 (*R*)
3	1c	20	24	5.2 (*R*)
4[b]	1d	20	>99	87.8 (*S*)
5	1e	20	>99	94.5 (*R*)
6	1b	-10	>99	98.2 (*R*)
7	1e	-10	>99	96.2 (*R*)

[a]General conditions: substrate/catalyst= 1000/1; t= 20 h; ligand/rhodium= 1/1; solvent= CH_2Cl_2. [b] Substrate/catalyst= 250/1.

The results indicated that the sense of enantiodiscrimination is predominantly controlled by the configuration of the binaphthyl moiety (entries

4 and 5). Moreover, they observed a cooperative effect between the stereogenic centers of the ligand backbone and the stereogenic binaphthyl phosphite moieties (entries 4 and 5). This resulted in a matched combination for ligand **1e** (entry 5 and 7).

They also found that ligand **1b** with conformational flexibility in readily epimerizing biphenyl moieties proved superior to those with fixed binaphthyl chirality (entry 2 vs 4 and 5).

The results obtained in the asymmetric hydrogenation of methyl *N*-acetamidoacrylate followed the same trend as those for dimethyl itaconate, but the enantioselectivities were somewhat lower (up to 88.7% ee).

The group of Claver and coworkers have recently developed a series of highly modular C1-diphosphite ligands with furanoside backbone derived from D-xylose and D-glucose (Figure 4) (20,21,22). The modular construction of these ligands allows sufficient flexibility to fine-tune (a) the different configurations of the carbohydrate backbone and (b) the steric and electronic properties of the diphosphite substituents.

Both excellent enantioselectivities (ee up to >99%) and activities were achieved in the Rh-catalyzed hydrogenation of dimethyl itaconate (Table 2). Systematic variation of stereocenters C-3 and C-5 at the ligand backbone revealed that enantiomeric excesses depended strongly on the absolute configuration of C-3 and slightly on that of the stereocenter carbon C-5. Therefore, enantioselectivities were best with ligands **4** with *R* configuration on both C-3 and C-5 stereocenters.

Variation in chirality at the axial chiral binaphthyl substituents in ligands **4** indicates that the sense of the enantiodiscrimination is predominantly controlled by the configuration of the biaryls at the phosphite moieties (entries 6 and 7). The presence of bulky substituents at the *ortho*-positions of the biaryl diphosphite moieties has a positive effect on enantioselectivity. The highest enantiomeric excess was found for allofuranoside ligand **4g**, which has *o*-trimethylsilyl substituents in the biphenyl moieties (entries 5 and 13).

It was also found that the presence of a methyl substituent on the carbon C-5 significantly increased the activity (entries 3-12 vs 1 and 2).

This set of ligands was also applied in the Rh-catalyzed hydrogenation of other benchmark dehydroamino acid derivatives. The results followed the same trend as those for dimethyl itaconate, but the activities were somewhat higher (21).

Monophosphite ligands

In the last few decades it was generally accepted that enantioselective hydrogenation was more effective in the presence of bidentate ligands. Recently, however, some monophosphorous ligands have been found to be very efficient for the Rh-catalyzed asymmetric hydrogenation (23,24,25). In the last

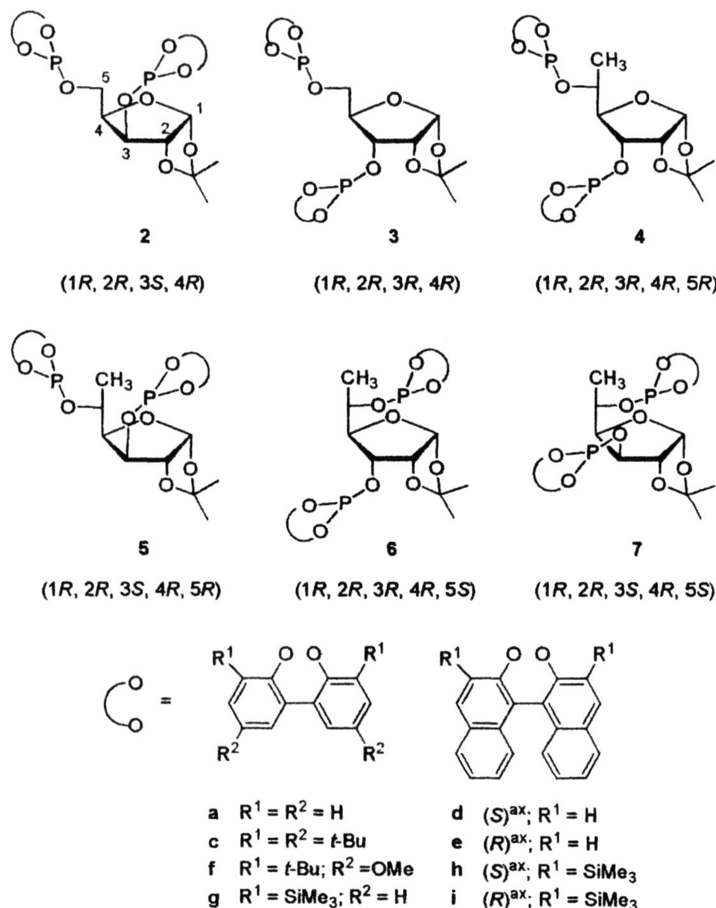

Figure 4. Diphosphite ligands **2-7** with furanoside backbone.

Table 2. Rh-catalyzed asymmetric hydrogenation of dimethyl itaconate using diphosphites 2-7.[a]

$$MeO_2C{-}C(=CH_2){-}CH_2{-}C(=O){-}Me \xrightarrow[\text{Rh / 2-7}]{H_2} MeO_2C{-}C^{*}H(CH_3){-}CH_2{-}C(=O){-}Me$$

Entry	*Ligand*	*% Conv (t/h)*	*% ee*
1	**2c**	12 (8)	22 (*R*)
2	**3c**	28 (8)	64 (*R*)
3	**4c**	90 (8)	90 (*R*)
4	**4f**	82 (8)	85 (*R*)
5	**4g**	100 (6)	97 (*R*)
6	**4d**	50 (8)	50 (*S*)
7	**4e**	46 (8)	52 (*R*)
8	**4h**	100 (8)	90 (*S*)
9	**4i**	100 (8)	92 (*R*)
10	**5c**	100 (8)	2 (*R*)
11	**6c**	87 (8)	67 (*R*)
12	**7c**	73 (8)	29 (*R*)
13[b]	**4g**	100 (4)	> 99 (*R*)

[a] $[Rh(cod)_2]BF_4$ = 0.01 mmol; ligand/Rh = 1.1; substrate/Rh = 100; CH_2Cl_2 = 6 mL; P_{H2} = 5 bar; T = 25 °C. [b] T = 5 °C; P_{H2} = 30 bar

few years, therefore, the use of monophosphite ligands in this process has been widely studied.

Research in this area was initiated by Reetz and coworkers (26). In connection with the previously described diphosphite ligands derived from mannitol **1**, they found out that the related monophosphite ligands **8d** and **8e** provided similar enantioselectivities (Figure 5a), so they extended their study to other simple binol-based monophosphite ligands **9** (Figure 5b) and found excellent results when an appropriate phosphite substituent was choosen. The results (100% conversion and 98.8% ee) were therefore best with ligands **9p** and **9q** containing chiral 1-phenylethanol units. However, the additional stereogenic center appears to play a minor role since the results were the same for both ligands.

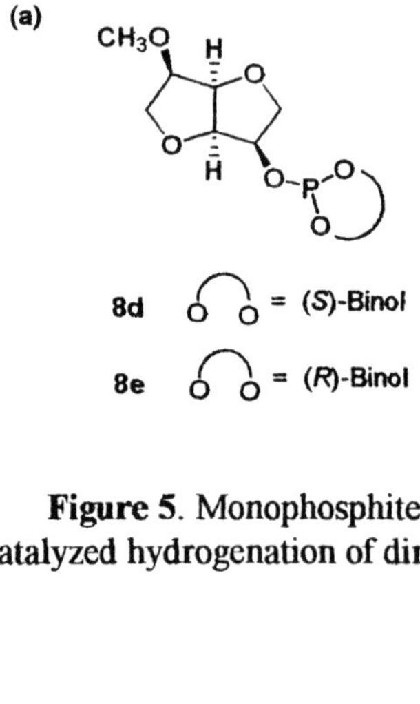

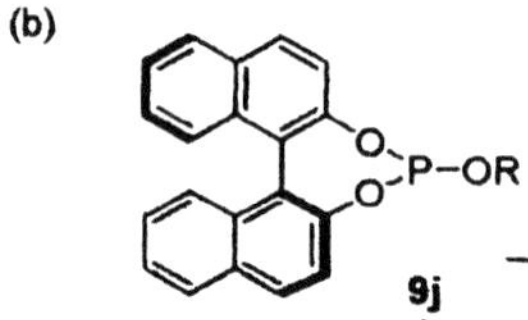

	R	ee
9j	CH_3	89.2%
k	$CH(CH_3)_2$	97.6%
l	C_6H_5	96.6%
m	2-Br-C_6H_4	89.8%
n	2,6-CH_3-C_6H_3	39.2%
o	2,6-(C_6H_5)-C_6H_3	28.6%
p	(*R*)-$CH(CH_3)(C_6H_5)$	99.2%
q	(*S*)-$CH(CH_3)(C_6H_5)$	98.2%

Figure 5. Monophosphite ligands tested by Reetz and coworkers in the Rh-catalyzed hydrogenation of dimethyl itaconate.

This set of ligands was also applied in the Rh-catalyzed hydrogenation of methyl *N*-acetamidoacrylate. The results followed the same trend as those for dimethyl itaconate, but the enantioselectivities were somewhat lower (ee's up to 95.5%).

Reetz and coworkers have taken advantage of these highly modular ligands **9** to show that catalyst optimization for a given substrate is possible simply by varying the alcohol unit (R) incorporated to the binaphthol basic framework. Enantioselectivities were therefore excellent in the asymmetric hydrogenation of different enamides using these highly modular ligands (Figure 6) (27).

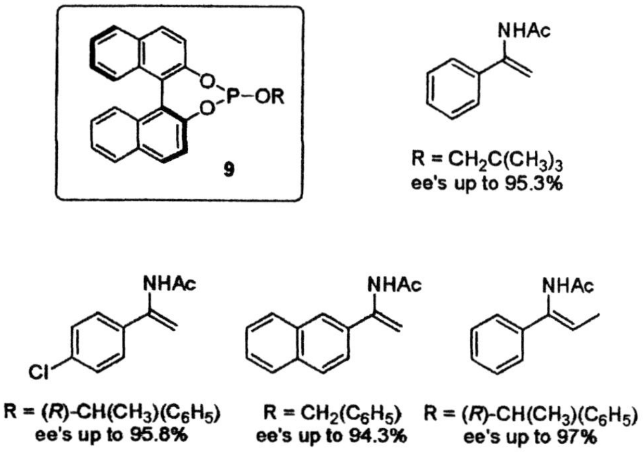

9

$R = CH_2C(CH_3)_3$
ee's up to 95.3%

$R = (R)-CH(CH_3)(C_6H_5)$
ee's up to 95.8%

$R = CH_2(C_6H_5)$
ee's up to 94.3%

$R = (R)-CH(CH_3)(C_6H_5)$
ee's up to 97%

Figure 6. Optimized enantioselectivities obtained in the Rh-catalyzed asymmetric hydrogenation of various enamides using ligands **9**.

Other groups have also explored the use of this type of ligand for the asymmetric hydrogenation of carbon-carbon double bonds (28, 29, 30). In particular, Dreisbach and coworkers have recently extended the study to enantiomerically pure biphenyl-based monophosphites (Figure 7). They studied the effect of the substitutents (R_1-R_4) of the biphenyl moiety in combination with the variation of the alcohol unit attached to the biphenol basic framework (R) and found excellent ligands for the rhodium-catalyzed asymmetric hydrogenation of dimethyl itaconate and methyl *N*-acetamidoacrylate (ee's up to 99%) (30).

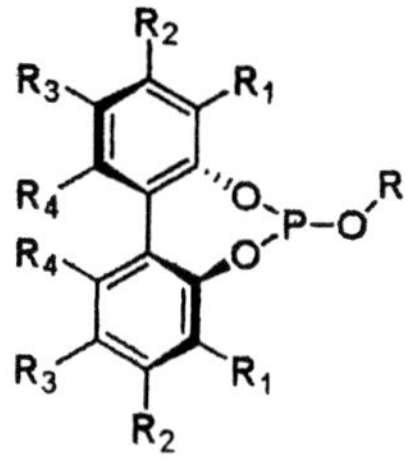

Figure 7. General structure of the biphenyl monophosphite ligands developed by Dreisbach and coworkers.

Very recently, Reetz and coworkers have developed a new concept in combinatorial asymmetric catalysis (31). These authors discovered that certain mixtures of different chiral monodentate ligands (in which two such ligands are bonded to rhodium) are much more active and enantioselective than either of the pure ligand systems (Scheme 2).

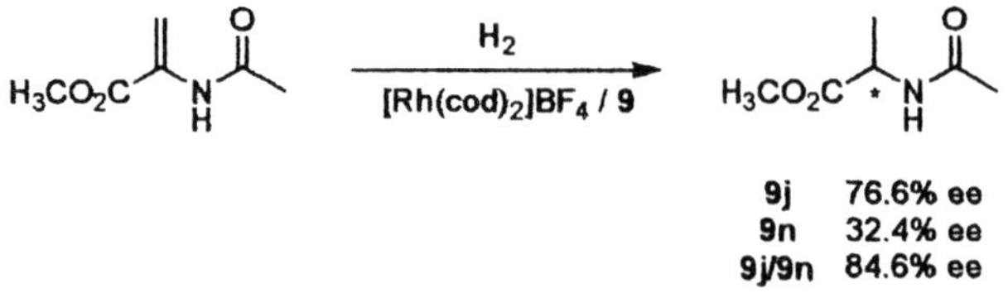

9j	76.6% ee
9n	32.4% ee
9j/9n	84.6% ee

Scheme 2. Comparison of the traditional and the combinatorial approach in the Rh-catalyzed asymmetric hydrogenation.

Hydrogenation of carbon-nitrogen double bonds

The enantioselective hydrogenation of carbon-nitrogen double bonds is a simple and convenient route to the synthesis of chiral amines. However, while many highly enantioselective chiral catalysts have been developed for the asymmetric hydrogenation of C=C and C=O, only very few effective catalysts are available for the enantioselective hydrogenation of C=N (1-3). The most widely used catalyst systems for this process are the diphosphine-rhodium (I) and diphosphine-iridium (I) complexes.

Little attention has been paid to the use of phosphite ligands for the asymmetric hydrogenation of C=N. So far only two reports are known in the literature (32, 33). The first of these reports used a binol-based diphosphite ligand **10** in the Ir-catalyzed asymmetric hydrogenation of unhindered imines and obtained low enantioselectivies (<12 % ee) (Scheme 3) (32).

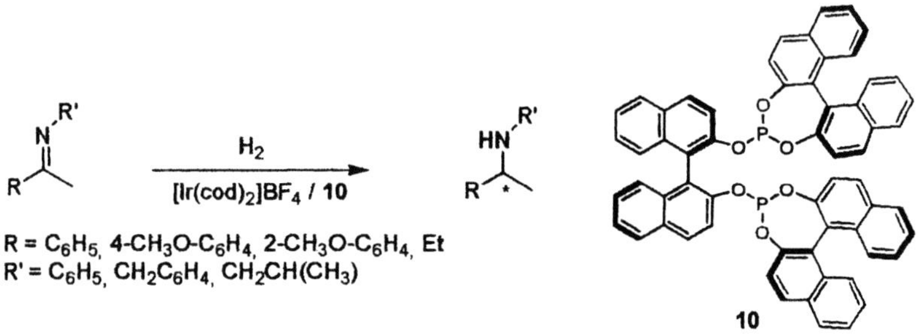

R = C_6H_5, 4-CH_3O-C_6H_4, 2-CH_3O-C_6H_4, Et
R' = C_6H_5, $CH_2C_6H_4$, $CH_2CH(CH_3)$

Scheme 3. Rh-catalyzed asymmetric hydrogenation of unhindered imines.

The second of these reports applied the previously described xylose-based diphosphite ligands (**2a**, **2c**) in the Ir-catalyzed asymmetric reduction of *N*-(phenylethylidene)aniline and obtained poor-to-moderate enantioselectivities (up to 46%) (33). The results indicated that enantioselectivity is strongly affected by the use of additives and the substituents in the biphenyl moiety (Table 3).

Table 3. Ir-catalyzed asymmetric hydrogenation of *N*-(phenylethylidene)aniline using diphosphites 2a and 2c.[a]

H_2

$[Ir(cod)_2]BF_4$/ 2

Entry	*Ligand*	*Additive*	*% Conv*	*% ee*
1	**2a**	-	40	0
2	**2c**	-	100	5
3	**2c**	Bu_4NI	100	46
4	**2c**	I_2	100	31
5	**2c**	Phthalimide	47	9
6	**2c**	$BnNH_2$	7.2	7.2

[a] $[Ir(cod)_2]BF_4$ = 1 %; ligand/Ir = 1; time = 16 h; solvent = CH_2Cl_2; P_{H2} = 70 bar; T = 25 °C.

Conclusions

During the last few years phosphite has becomes one of the most versatile ligands for the enantioselective hydrogenation reaction. Excellent control of substrate selectivity based on the properties of the ligand has been demonstrated. For industrial applications, however, the productivity of the catalysts needs to be further improved in order to achieve high turnover numbers and frequencies. Its is to hope that the efficiency of the catalyst can be increased and that the search for improved ligand systems will be greatly assisted by combinatorial screening methods, to which the readily available phosphite ligand may prove to be well suited.

References

1. *Asymmetric Catalysis in Organic Synthesis*; Noyori, R., Ed.; Wiley, New York, 1994.
2. *Catalytic Asymmetric Synthesis*; Ojima, I., Ed.; Wiley, New York, 2000.
3. *Comprehensive Asymmetric Catalysis*; Jacobsen, E. N.; Pfaltz, A.; Yamamoto, H., Eds; Springer, Berlin, 1999, vol 1.
4. Babin, J. E.; Whiteker, G. T. (Union Carbide Chem. Plastics Techn. Co.) WO 93/03839, 1993 [Chem. Abs. 1993, 119, P159872h]
5. Buisman, G. J. H. ; van deer Veen, L. A.; Klootwijk, de Lange, W. G. J. ; Kamer, P. C. J.; van Leeuwen, P. W. N. M.; Vogt, D. *Organometallics* **1997**, *16*, 2929.
6. Nozaki, K.; Sakai, N.; Nanno, T.; Higashijima, T.; Mano, S.; Horiuchi, T.; Takaya, H. *J. Am. Chem. Soc.* **1997**, *119*, 4413
7. Franciò, G.; Leitner, W. *Chem. Commun.* **1999**, 1663
8. Diéguez, M.; Pàmies, O.; Ruiz, A.; Castillón, S.; Claver, C. *Chem. Eur. J.* **2001**, *7*, 3086
9. *Rhodium Catalyzed Hydroformylation*; van Leeuwen, P. W. N. M.; Claver, C. Eds.; Kluwer Academic Press, Dordrecht, 2000
10. Horiuchi, T.; Shirakawa, E.; Nozaki, K.; Takaya, H. *Tetrahedron: Asymmetry* **1997**, *8*, 57
11. Prétôt, R.; Pfaltz, A. *Angew. Chem. Int. Ed.* **1998**, *37*, 323.
12. Diéguez, M.; Jansat, S.; Gomez, M.; Ruiz, A.; Muller, G.; Claver, C. *Chem. Commun.* **2001**, 1132
13. Selvakumar, K.; Valentini, M.; Pregosin P. S.; Albinati, A. *Organometallics* **1999**, *18*, 4591
14. Pàmies, O.; van Strijdonck, G. P. F.; Diéguez, M.; Deerenberg, S.; Net, G.; Ruiz, A.; Claver, C.; Kamer, P. C. J.; van Leeuwen, P. W. N. M. *J. Org. Chem.* **2001**, *66*, 8867
15. Brunner, H.; Pieronczyk, W. *J. Chem. Res (S)* **1980**, 76.
16. Wink, D. J.; Kwok, T. J.; Yee, A. *Inorg. Chem.* **1990**, *29*, 5006.
17. Kolich, C. H. US Patent 5210243, 1993
18. Kadyrov, R.; Heller, R.; Selke, R. *Tetrahedron: Asymmetry* **1998**, *9*, 329.
19. Reetz, M. T.; Neugebauer, T. *Angew. Chem. Int. Ed.* **1999**, *38*, 179.
20. Pàmies, O.; Net, G.; Ruiz, A.; Claver, C. *Eur. J. Inorg. Chem.* **2000**, 1287.
21. Diéguez, M.; Ruiz, A.; Claver, C. *J. Org. Chem.* **2002**, *67*, 3796.
22. Diéguez, M.; Ruiz, A.; Claver, C. *Dalton Trans.*, 2003 in press.
23. Claver, C.; Fernandez, E.; Gillon, A.; Heslop, K.; Hyett, D. J.; Martorell, A.; Orpen, A. G.; Pringle, P. G. *Chem. Commun.* **2000**, 961.
24. Reetz, M. T.; Mehler, G.; Meiswinkel, A.; Sell, T. *Tetrahedron Lett.* **2002**, *43*, 7941.

25. Peña, D.; Minnaard, A. J.; de Vries, J. G.; Feringa, B. L. *J. Am. Chem. Soc.* **2002**, *124*, 14552.
26. Reetz, M. T.; Mehler, G. *Angew. Chem. Int. Ed.* **2000**, *39*, 3889.
27. Reetz, M. T.; Mehler, G.; Meiswinkel, A.; Sell, T. *Tetrahedron Lett.* **2002**, *43*, 7941.
28. Chen. W.; Xiao, J. *Tetrahedron Lett.* **2001**, *42*, 2897.
29. Chen. W.; Xiao, J. *Tetrahedron Lett.* **2001**, *42*, 8737.
30. Dreisbach, C.; Meseguer, B.; Prinz, T.; Scholz, U.; Militzer, H. C.; Agel, F.; Driessen-Hoelscher, B. *Eur. Pat. Appl.* 1298136 A2, 2003.
31. Reetz, M. T.; Sell, T.; Meiswinkel, A.; Mehler, G. *Angew Chem. Int. Ed.* **2003**, *42*, 790.
32. Martorell, A.; Claver, C.; Fernández, E. *Inorg. Chem. Comun.* **2000**, *3*, 132.
33. Guiu, E.; Muñoz, B.; Castillón, S.; Claver, C. *Adv. Synth. Catal.* **2003**, *345*, 169.

Chapter 12

Amino Acid-Based Phosphorous(III) Ligands

D. Jayasinghe and H. B. Kraatz

Department of Chemistry, University of Saskatchewan, 110 Science Place, Saskatoon, Saskatchewan S7N 5C9, Canada

The synthesis of chiral ligands using naturally occurring and renewable sources of chirality has become an important area of research. Chiral phosphorous ligands based on amino acids can be readily synthesized in a few steps giving useful components to catalytic systems. Here we provide a brief overview of current ligands and applications.

Phosphorous-based ligands have attracted tremendous attention and are among the most important ligands in organometallic chemistry with a wide range of steric and electronic properties. In particular, for asymmetric transformations these chiral phosphine ligands have found wide-spread applications[1-3]. However, their synthesis often involves multistep reactions. Chiral biomolecules on the other hand, such as amino acids, peptide and carbohydrates, offer a particularly convenient starting point for the synthetis of chiral phosphorous ligands and are readily available in an enantiopure form. In addition, they are ready accessible, renewable and often of low cost. Carbohydrates, have been exploited as chiral starting materials for the synthesis of phosphinites, making use of the rich array of stereochemical and functional group diversity[4-9]. These have found applications in catalysis[4,6]. However, in some cases, the desired enantiomer for a particular reaction leading to a specific enantiopure product may not be available from natural sources and thus maybe very costly. Amino acids, on the other hand, are readily available in both enantiomeric forms (*L*-phenylalanine CA$ 1.25/g; *D*-phenylalanine CA$ 6.75/g; Aldrich-2003/2004) when compared to that of sugars, (αD-glucose CA$ 0.98/g; L-glucose CA$ 106.8/g; Aldrich- 2003/2004) and may facilitate the synthesis of both enantiomers of the intended product. Amino acids and their derivatives have also found widespread applications. In particular, proline-derived aminoalcohols have been popularized by Mortreux and coworkers for the synthesis of chelating aminophosphine-phosphinite ligands [10-13]. More recently, amino acid and peptide-based phosphines and phosphinites have been reported in the literature.

Here we wish to summarize recent results on the synthetic aspects of amino acid and peptide-based phosphinite ($R_2P(OR)$) and phosphine (R_2PR) ligands and highlight some of their applications.

Phosphinites

Phosphinites provide different chemical, electronic and structural properties compared to phosphines. As was shown by RajanBabu and others, phosphinites are versatile ligands which allow effective catalytic asymmetric transformations. The metal-phosphorous bond is often stronger in phosphinites compared to the related phosphines due to the presence of the electron-withdrawing P-OR group. In addition, the empty σ*-orbital of the phosphinite, $P(OR)R_2$ is stabilized and thus a better acceptor. The synthesis of phosphinites is conveniently achieved by reacting alcohols with dialkyl or diarylphosphinechlorides in the presence of a base. Using this approach, Mortreux was able to synthesize chiral amino alcohol based aminophosphine-phosphinite (AMPP) ligands. He proposed two parallel synthetic pathways to generate chiral AMPP ligands from amino acids, either by direct synthesis from amino alcohols (Scheme 1), or by the reduction of formyl esters of α-amino acids followed by coupling with PPh_2Cl (Scheme 2) [12].

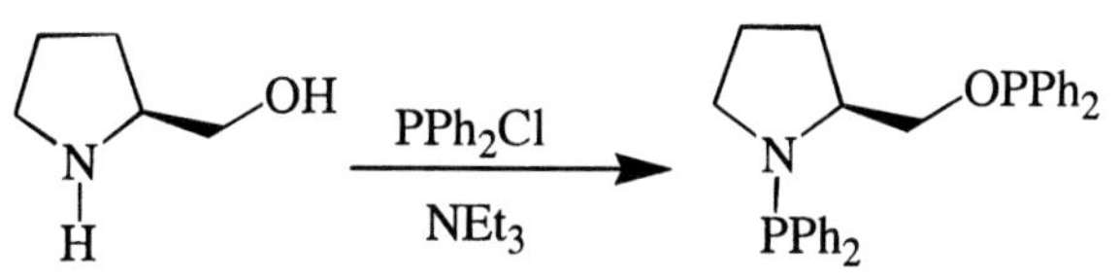

PPh$_2$Cl, NEt$_3$

OH → OPPh$_2$; N–H → N–PPh$_2$

Scheme 1: *Synthesis of a prolinol-based aminophosphine-phosphinite (AMPP) ligands*

In contrast to this, recent work from our group has shown the use of protected hydroxyl-containing amino acids to form phosphinites and even peptido-phosphinites.

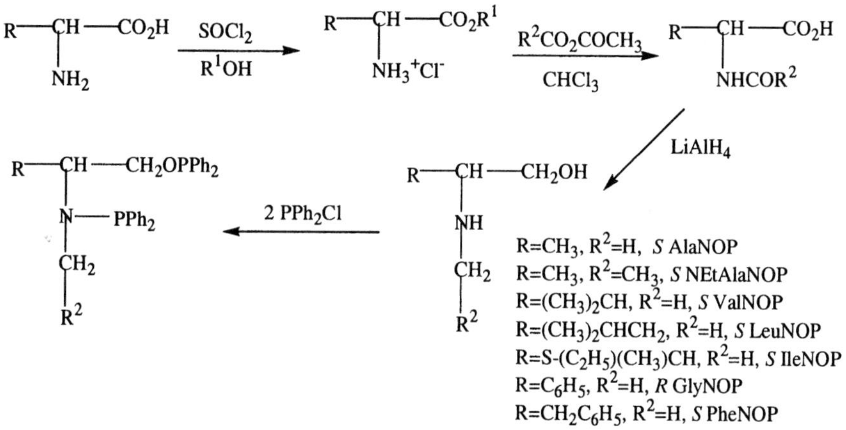

R—CH(NH$_2$)—CO$_2$H → (SOCl$_2$, R^{1}OH) → R—CH(NH$_3^+$Cl$^-$)—CO$_2$R^1 → (R^{2}CO$_2$COCH$_3$, CHCl$_3$) → R—CH(NHCOR^2)—CO$_2$H → (LiAlH$_4$) → R—CH(NH—CH$_2$—R^2)—CH$_2$OH → (2 PPh$_2$Cl) → R—CH(N(PPh$_2$)—CH$_2$—R^2)—CH$_2$OPPh$_2$

R=CH$_3$, R^2=H, *S* AlaNOP
R=CH$_3$, R^2=CH$_3$, *S* NEtAlaNOP
R=(CH$_3$)$_2$CH, R^2=H, *S* ValNOP
R=(CH$_3$)$_2$CHCH$_2$, R^2=H, *S* LeuNOP
R=S-(C$_2$H$_5$)(CH$_3$)CH, R^2=H, *S* IleNOP
R=C$_6$H$_5$, R^2=H, *R* GlyNOP
R=CH$_2$C$_6$H$_5$, R^2=H, *S* PheNOP

Scheme 2: *General synthetic scheme for the synthesis of amino alcohol-based aminophosphine-phosphinite (AMPP) ligands from amino acids*

The reaction of amino acids, such as serine (**1**), threonine, and tyrosine, with one equivalent of chlorodiphenylphosphines in the presence of an organic base gives the corresponding phosphinites in good yields. (Scheme 3)[14].

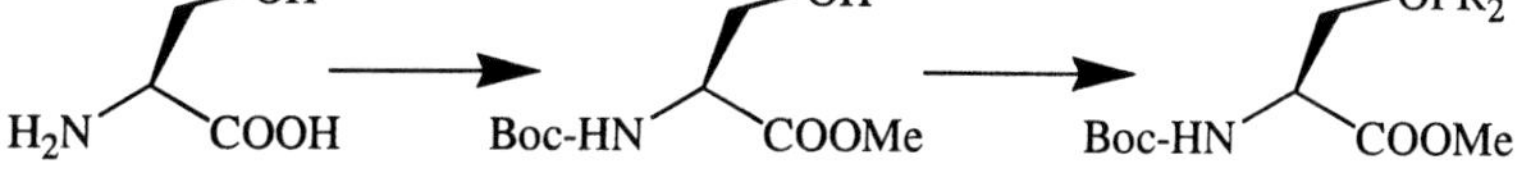

H$_2$N–CH(CH$_2$OH)–COOH (**1**) → Boc-HN–CH(CH$_2$OH)–COOMe (**2**) → Boc-HN–CH(CH$_2$OPR$_2$)–COOMe (**3**)

Scheme 3: *General synthesis of Boc-Ser(OPR$_2$)-OMe (3) (R = Ph) from serine (1) via the N,C-protected serine derivative Boc-Ser(OH)-OMe (2).*

The corresponding phenylphosphinites of threonine and tyrosine (**4** and **5**) are readily prepared under identical conditions in high yields. Other commonly used N-protecting groups do not interfere with phosphinite formation and allow the formation of e.g. Z-Ser($OPPh_2$)-OMe (**6**). This strategy is readily extended to include the preparation of previously unknown peptide-phosphinites, such as Z-β-Ala-Tyr($OPPh_2$)-OMe (**7**).

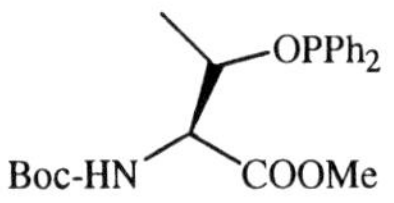

Boc-HN COOMe OPPh2 **4** Boc-HN COOMe OPPh2 **5** Z-HN COOMe OPPh2 **6**

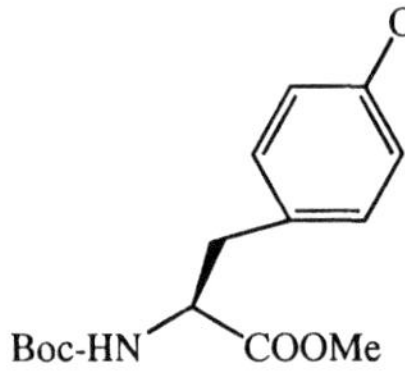

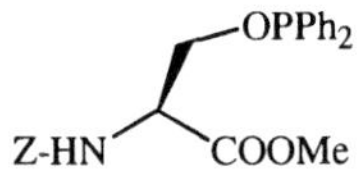

Z-HN O N H COOMe OPPh2 **7**

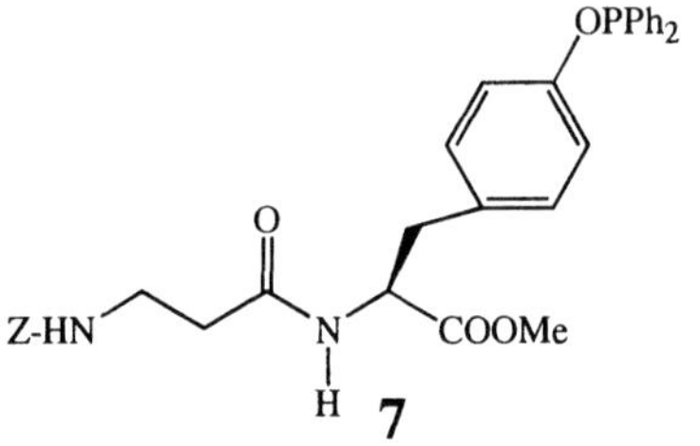

This may prove a useful strategy for the generation of chiral phosphorous ligands in which the structure of the peptide backbone may influence the outcome of specific catalytic reactions by generating a chiral environment encapsulating the metal reaction site.

A number of metal complexes of AMPP and amino acid phosphinites are known. In particular Pd(II) and Pt(II) complexes of the phosphinite ligands **3** - **7** were prepared in quantitative yields by reacting the appropriate phosphinite with

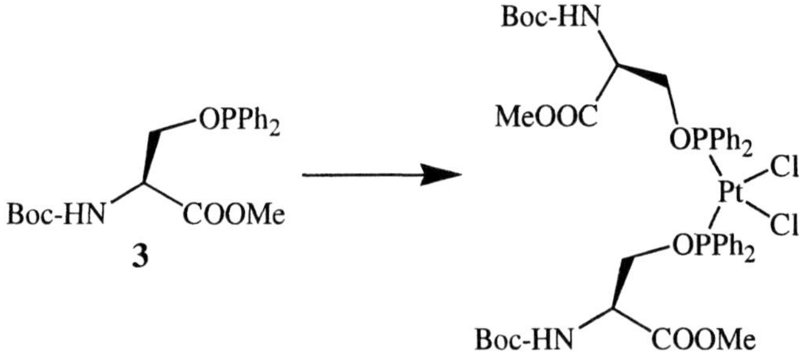

Boc-HN COOMe OPPh2 **3** → Boc-HN MeOOC OPPh2 Pt Cl Cl OPPh2 Boc-HN COOMe

Scheme 4: *Synthesis of [Boc-Ser($OPPh_2$)-OMe]$_2$$PtCl_2$ as a general example for the preparation of metal phosphinite complexes*

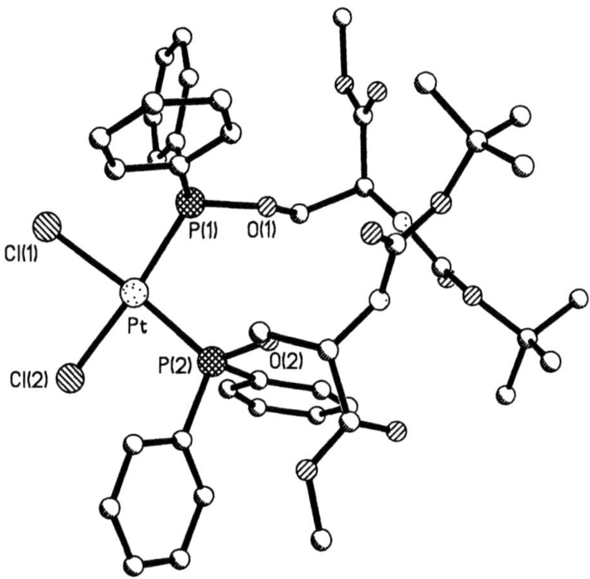

Figure 1: *Drawing of the molecular structure of the Ser-phosphinite complex* [Boc-Ser($OPPh_2$)-OMe]$_2$$PtCl_2$ *showing the coordination geometry.*

Table 1. *Selected $^{31}P\{^1H\}$ NMR chemical shift data of the amino acid phosphinito ligands (in $CDCl_3$, δ in ppm, J_{Pt-P} in Hz)*

	$^{31}P\{1H\}$		
Ligand L		Cl_2PdL_2	Cl_2PtL_2
BocSer($OPPh_2$)OMe (**3**)	118.3	112.6	84.8 (4100)
BocThr($OPPh_2$)OMe (**4**)	113.3	103.9	76.4 (4100)
BocTyr($OPPh_2$)OMe (**5**)	111.1	111.7	86.1 (4220)
ZSer($OPPh_2$)OMe (**6**)	118.3		
ZβAla-Tyr($OPPh_2$)OMe (**7**)	111.5	111.5	86.0 (3760)

either Pd or Pt precursor as outlined in Scheme 4. The structure of the Pt(II)-Ser-phosphinite complex is shown in Figure 1. The complex exhibits the typical square planar coordination environment around the Pt(II) center, having both phosphinites in a *cis* coordination. This is compatible with the spectroscopy of Pd(II) and Pt(II) complexes. All metal complexes exhibit signals in the ^{31}P-NMR spectrum that are upfield from the position of the free ligand (Table 1). The ^{31}P-^{195}Pt coupling constants and the IR spectra agree with a *cis* arrangement of the phosphinite ligands around the metal center.

Phosphines

Phosphines have proven extremely versatile ligands for a wide range of transition metal catalyzed transformations [1-3]. Chiral phosphines, such as the bidentate DIOP, DUPHOS, and others coordinated to late transition metals have allowed asymmetric transformations with high yields and high ee's [15-16]. Inexpensive optically pure amino acids have been converted to both enantiomers and have been employed in the laboratory in numerous asymmetric catalytic systems as will be described later.

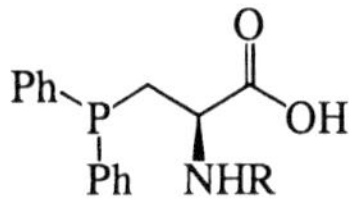

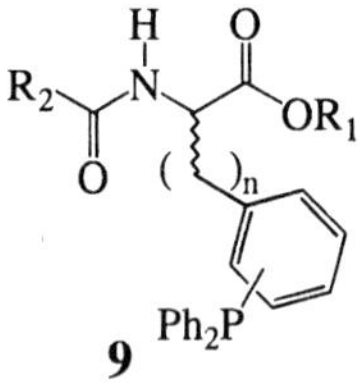

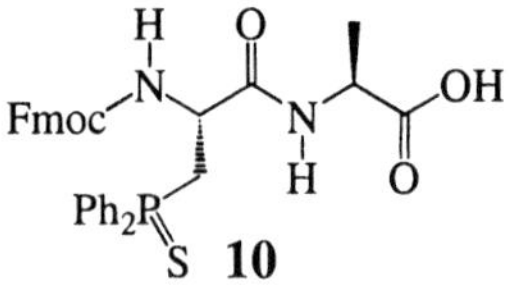

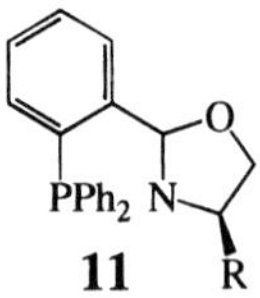

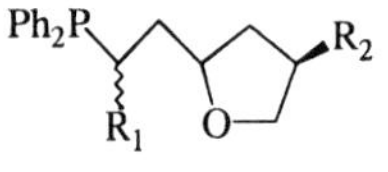

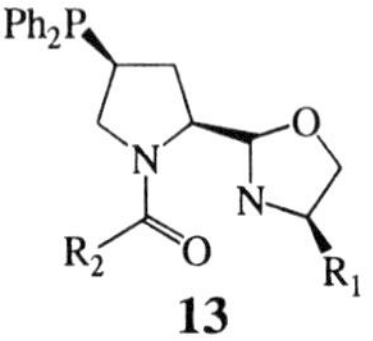

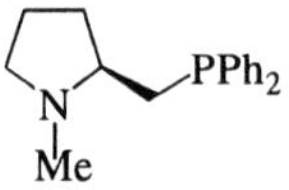

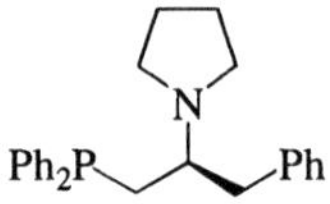

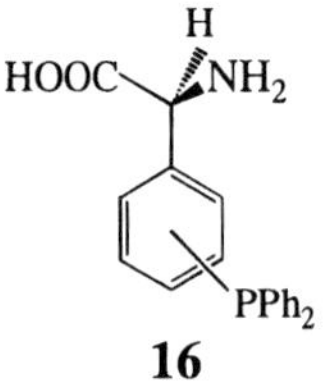

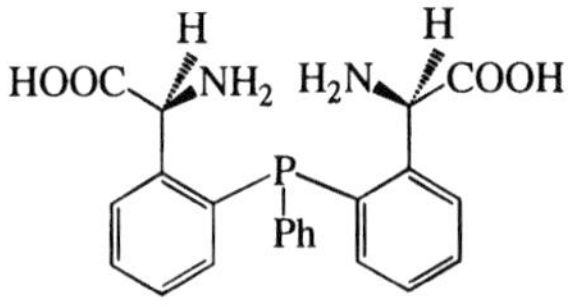

Recently Gilbertson and others have described the synthesis of a number of chiral phosphine ligands (**8** - **15**)[17-27] , **(16-17)** [29]. Pd(0)-catalyzed P-C bond forming reactions are particularly useful to make these ligands. However, in reactions involving aryl triflates with diphenylphosphine-borane adducts racemization of the amino acid was observed. However, cross-coupling of C,N-protected iodophenylalanine using triethylamine as a base, it was possible to obtain *N*-acetyl-(4-diphenylphosphinanyl-phenyl)-alaninemethylester directly without any noticeable racemization at the α-carbon. In addition, these conditions do not require a phosphine protection-deprotection sequence (Scheme 5) [29].

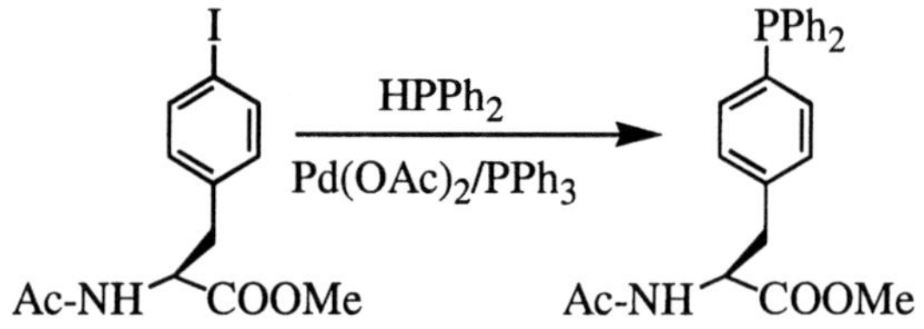

I | Ac-NH COOMe — $HPPh_2$ / $Pd(OAc)_2/PPh_3$ → PPh_2 | Ac-NH COOMe

Scheme 5: *Synthesis of L-PhePhos from 4-iodophenylalanine*

Glibertson reported the synthesis and metal coordination chemistry of phosphino peptides. This work was followed by a report of the synthesis of a novel library of phosphine ligands and phosphino oxazoline ligands (22) starting from serine or threonine derivatives. Importantly all reactions involve a P-protection/deproection sequence to avoid oxide formation. This work was extended to the synthesis of phosphine containing peptide libraries using phosphine containing serine as building block. These ligands were successfully tested in asymmetric hydrogenation studies[28]

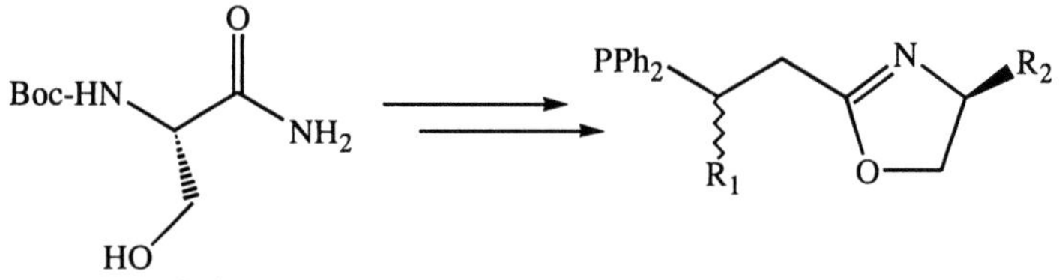

Boc-HN, O, NH_2, HO → PPh_2, N, R_2, R_1, O

Scheme 6: *Synthesis of phosphino-oxazoline ligands*

Stelzer et.al reported a useful phosphine synthesis starting from fluoroalanine in DMF. Nucleophilic substitution of diarylphosphide on the aromatic group results in the formation of the unprotected amino acid. This pathway may prove extremely useful for the incorporation of phosphino amino acids into larger peptide sequences (29).

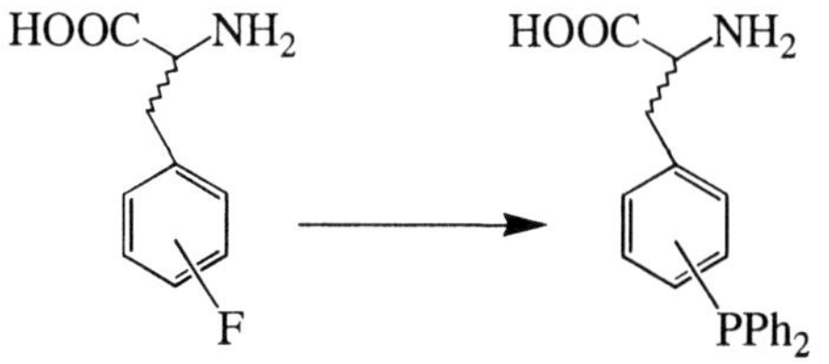

HOOC NH2 F → HOOC NH2 PPh2

Scheme 7: Formation of diphenylphosphino phenylalanine

Catalytic Studies.

Early studies by Tzschach and corworkers show the utility of bidentate N-phosphine substituted alanine derivtives of the type $(Ph_2PCH_2)_2NCH(CH_3)COONa$ in the asymmetric hydrogenation of α-acetamidocinnamic acid derivatives activated olefins using $[Rh(1,5\text{-hexadiene})Cl]_2$ as the metal starting material (Scheme 8). However, the enantiomeric excess of the (+)-*S* products are low with (~ 30% ee) (31).

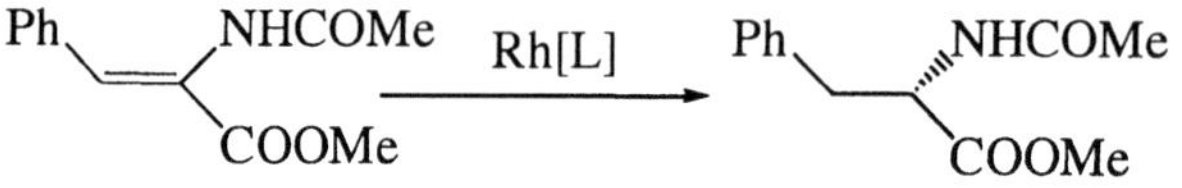

Ph NHCOMe COOMe —Rh[L]→ Ph NHCOMe COOMe

Scheme 8: Asymmetric hydrogenation of α-acetamidocinnamic acid derivatives

Morimoto and coworkers used bidentate N,P-chelats derived from amino acids as ligands (Scheme 9) in the Pd catalyzed asymmetric allylic substitution of acylic and cyclic pivalates, such as 1,3-diphenylpropen-2-yl pivalate and cyclohexen-2-yl pivalate (Scheme 10) (32).

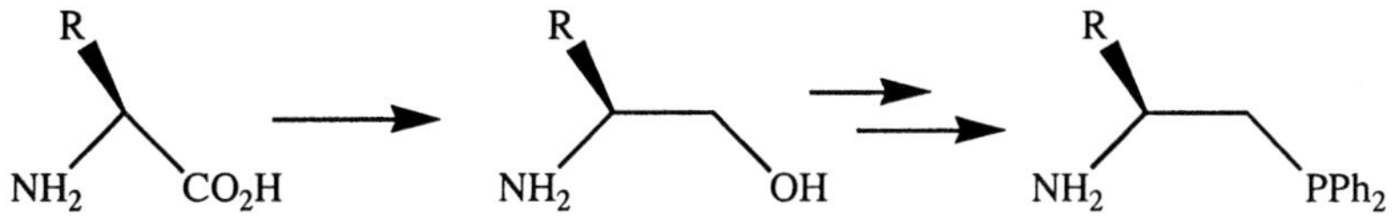

Scheme 9: Formation of amino acid derived N,P-chelates

These ligands provide excellent chiral induction and acyclic substrates are readily transformed into the desired ligands up to 95% ee (Scheme 10a). The transformation of cyclic substrates is particularly troublesome. However, the new N,P-ligand again proves its utility by providing more than 99% ee (Scheme 10b).

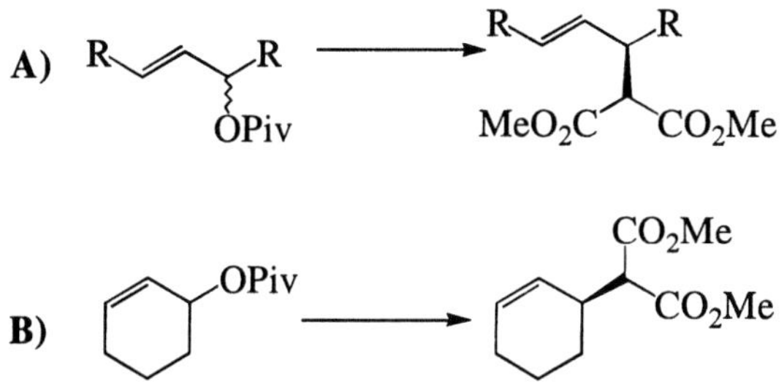

Scheme 10: Pd-catalyzed allylic substitution reactions for a) acylic and b) cyclic starting materials.

There are numerous examples of aminophosphine- and amidophosphine-phosphinito ligands based on hydroxyproline and its derivatives [34]. Gilbertson reported some catalytic studies using proline-based phosphine-oxazoline P,N-ligands (**13**) for Pd-catalyzed allylic alkylations and in Heck reactions [33].

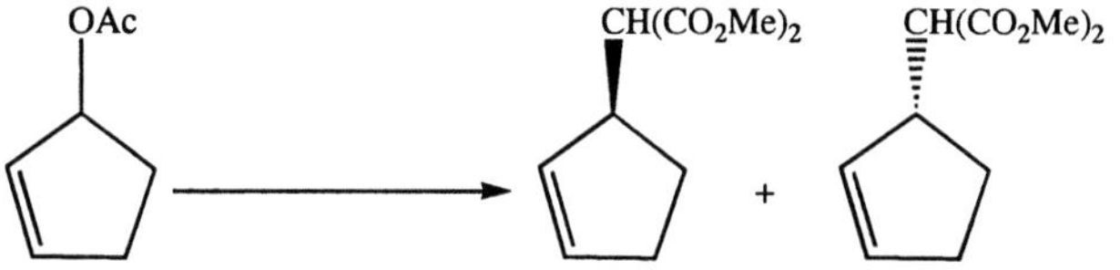

*Scheme 11: Pd-catalyzed allylic addition to cyclopentenyl acetate using ligand **13***

Ligand **13** appears very effective in controlling the addition of malonates to cyclic allyl acetates. Importantly, the substitutent at the proline nitrogen has a significant influence on the stereochemistry of the product. For example an Fmoc protected proline will result in the formation of 69%ee of *S*- cyclopentenyl malonate whereas tBuCO will result in the R isomer being formed in 35%ee.

Importantly, the stereochemistry of the ligand remains the same. This is rationalized by the ligand forming a binding pocket and the spatial requirement of the N-substituent.

The same ligand type was used for asymmetric Heck reactions of dihydrofuran and 1-cyclohexenyltriflate and aryl triflates (Scheme 12).

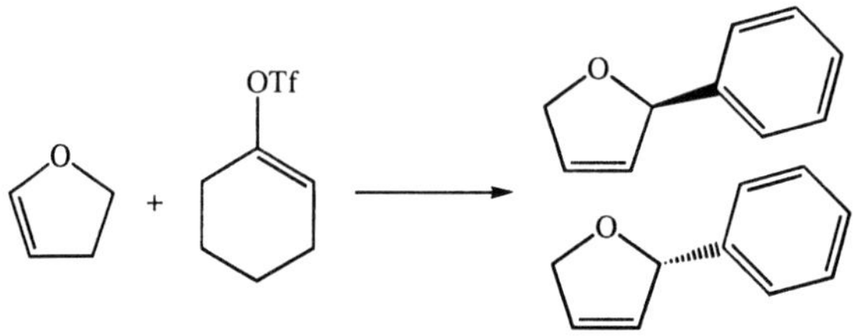

OTf

+

O

Scheme 12: Asymmetric Heck reactions involving dihydrofuran and cyclohexenyltriflate

The reaction gives high yields of the product with ee's ranging from 12 to 80%. High selectivities were observed with triethylamine and diisopropylethylamine as bases [33].

Ir(I) complexes of the aminophosphine proline-oxazoline ligands in which the proline-N carries a PPh_2 subsituent was recently reported for the asymmetric hydrogenation of methystilbenes resulting in product yields and good ee's for the corresponding *R*-alkane [37].

Some of proline-based ligands [35], were used for the hydrocarbonylation and hydrogenation of olefins. Agbssou and Morteux recently reported on the use of (1R,2S)-ephedrine based AMPP ligands for the asymmetric hydroformylation of styrene as shown in Scheme 13 [38]. Although a high region and chemoselectiviy towards the branched aldehyde was observed, the enatioselectivities were low.

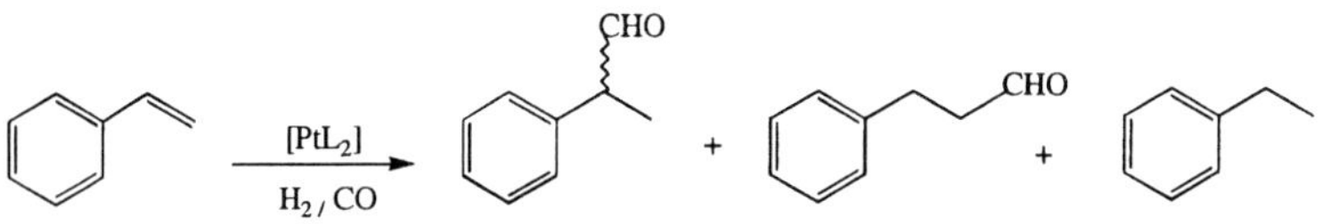

CHO

CHO

$[PtL_2]$

H_2/CO

+

+

Scheme 13: Hydrocarbonylation of styrene

Similarly disappointing results were reported for the hydrosilation of styrene using simple amino acid phosphinites [Ref 35]. Hydrogenation reactions using Rh(I) complexes of pyroglutammic acid derived AMPPs show good ee's as exemplified by the hydrogenation of geraniol and nerol to citronellol [37]. Rh(I) complexes of AMPPs derived from (*S*)-*N*-benzylmandelamide and its analogues give reasonable results for the asymmetric hydrogenation of dihydro-4,4-dimethyl-2,3-furandione and *N*-benzylbenzoylformamide. However, the ee's for the latter are relatively low [39]

Recently, Gilbertson reported examples of helical peptides carrying phosphino amino acid donors and described their catalytic behaviour [28]. Hydrogenation of methyl 2-acetamidoacrylate was performed using Rh(1) complexed to helical phosphinopeptides. The system is showing 99% ee for substrate and thus exhibits a greater selectivity than DuPHOS.

O OCH3 NHAc — H_2 / Rh^+ complexed to peptide on support → H_3C O OCH3 NHAc + H_3C O OCH3 NHAc

Acknowledgments

This research is supported by the National Science and Engineering Research Council of Canada. H.B.K is Canada Research Chair for Biomaterials.

References

(1) E. N. Jacobsen, A. Pfaltz, H. Yamamoto, Comprehensive Asymmetric Catalysis, Springer, New York, 1999, Vol. 1-3.
(2) H. B. Kagan, M. Sasaki, Transition Metals for Organic Synthesis, Wiley-VCH, Weinheim, 1998.
(3) R. Noyori, Asymmetric Catalysis in Organic Synthesis, Wiley, New York, 1994
(4) RajanBabu, T. V.; Ayers A. T.; Halliday G. A.; You, K. Y.; Calabrese, J. C.; *J. Org. Chem.* **1997**, 62, 6012-6028
(5) Yan, Y.; RajanBabu T. V.; *J. Org. Chem.* **2000**, 65, 900-906
(6) RajanBabu T. V.; Yan, Y.; Shin, S.; *J. Org. Chem* **2000**, 123, 10207-10213

(7) Yamashita, m.; Kobayashi, M.; Sugiura, M.; Tsunekawa, T.; Inokawa, S.; Yamamoto, H.; Bull. of the Chem. Soc. of Jpn. **1986**, 59(1), 175-8
(8) Selke, R.; Facklam, C.; Foken, H.; Heller, D. Tetrahedron Asymmetry **1993**, 4(3), 369-82
(9) Selke, R.; Reaction Kinetics and Catalysis Letters **1979**, 10(2), 135-8
(10) Karim, A.; Mottreux, A.; Petit, F. *Journal of Organometallic Chemistry* **1986**, 312(3), 375-81
(11) Naili, S.; Suisse, I.; Mortreux, A.; Agbossou-Niedercorn, F.; Journal of Molecilar Catalysis A: Chemical 164 (2000) 125-130
(12) Karim, A.; Mottreux, A.; Petit, F. .*J. of Organomet. Chem.* **1986**, 317, 93-104
(13) Ali, M. A.; Allaoud, S.; Karim, A.; Roucoux, A.; Mortreux, A.; Tetrahedron Asymmetry **1995**, Vol. 6, No.2, pp 369-370
(14) Galka, P. V.; Kraatz, H.B. *Jornal of Organometallic Chemistry* **2003**, 00, 1-8
(15) Robert, F.; Oehme, G.; Sinou, D.; J. of Mol. Catal. A: **1999**, 139, 105-108
(16) Toth, I.; Hanson B. E.; Davis, M. E. *Tetrahedron Asymmetry* **1990**, Vol. 1, No. 12, pp. 913-930
(17) Gilbertson S. R.; Chen. G.; McLoughlin, M. *J. Am. Chem. Soc.* **1994**, 116, 4481-4482
(18) Gilbertson S. R.; Wang, X. *J. Org. Chem.* **1996,** 61, 434-435
(19) Gilbertson S. R.; Starkey, G. W.. *J. Org. Chem.* **1996,** 61, 2922-2923
(20) Gilbertson S. R.; Wang, X. Hoge, G. S. Klung, C. A.; Schaefer, J. *Organometallics.* **1996,** 15, 6478-6480
(21) Gilbertson S. R.; Chen, G.; Kao, J.; Beatty, A.; Campana, C. F. *J. Org. Chem.* **1997,** 62, 5557-5566
(22) Gilbertson S. R.; Chang C. T. *J. Org. Chem.* **1998**, 63, 8424-8431
(23) Gilbertson S. R.; Xie, D.; Fu, Z. *J. Org. Chem.* **2000,** 66, 7240-7246
(24) Xu, G.; Gilbertson S. R. Tetrahedron Lett. **2002**, 43, 2811-2814
(25) Xu, G.; Gilbertson S. R. Tetrahedron Lett. **2003**, 44, 953-955
(26) Gilbertson S. R.; Wang, X. Tetrahedron 1999, 55, 11609-11618
(27) Gilbertson S. R.; Wang, X. *Tetrahedron Lett.* 1996, 37, 6475-6478
(28) Gilbertson S. R.; Wang, X. Tetrahedron **1999**, 55, 11609-11618
(29) Brauer, D. J.; Schenk. S.; RoBenbach. S.; Tepper, M.; Stelzer, O.; Hausler T.; Sheldrick, W. S. *J. of Organomet.Chem.* **2000**, 598, 116-126
(30) Kraatz, H. B.; Pletsch, A. Tetrahedron: *Asymmetry* **2000,** 11, 1-5
(31) Kellner, K.; Hanke, W.; Tzschach, A.; Nagy-Magos, Z.; Marko, L. *J. Organomet. Chem.* **1984**, *268*, 175.
(32) Saitoh, A.; Achiwa, K.; Tanaka, K.; Morimoto, T. *J. Org. Chem.* **2000,** *65*

(33) Gilbertson, S.R.; Xie, D.; Fu, Z. *J. Org. Chem.* **2001**, *66*, 7240.

(34) A. Roucoux, L. Thieffry, J.-F. Carpentier, M. Devocelle, C. Méliet, F. Agbossou and A. Mortreux, *Organometallics*, **15**, 2440 (1996), and references therein. b) F. Agbossou, J.-F. Carpentier and A. Mortreux, *Chem. Rev.* **95**, 2485 (1995); c) S. R. Gilbertson and D. Xie, *Angew. Chem. Int. Ed. Engl.* **38**, 2750 (1999) and references therein.

(35) M. Petit, A. Mortreux, F. Petit, G. Buono, G. Peiffer, New J. Chem. 7 (1983) 593. b) J. W. Scott, Top. Stereochem. 19 (1989) 209

(36) Galka, P.V.; Kraatz, H.B. in: Catalysis of Organic Reactions, M.E. Ford (Ed.), M. Dekker, New York, 2001, pp. 589-594

(37) Xu, G. ; Gilbertson, S.R. Tetrahedron Lett. **2003**, *44*, 953-955.

(38) Lot, O. ; Suisse, I. ; Mortreux, A.; Agbossou, *J. Mol. Catal.A* **2000**, *164*, 125.

(39) Roucoux, A. ; Thieffry, L.; Carpentier, J.-F.; Devocelle, M.; Mélier, C. ; Agbossou, F. ; Mortreux, A. *Organometallics* **1996,** *15*, 2240-2449.

Chapter 13

Asymmetric Catalysis with Chiral Lithium Amides

Sanjay V. Malhotra

Department of Chemistry and Environmental Science, New Jersey Institute of Technology, University Heights, Newark, NJ 07102–1982

Chiral lithium amides (CLAs) have become important tool for asymmetric transformations. Catalytic reactions using these bases have been studied for desymmetrization of prochiral epoxides and ketones. This chapter gives a brief overview of the recent developments in the catalytic reactions using chiral lithium amides.

Asymmetric synthesis using chiral lithium amides is emerging as a useful method for the preparation of non-racemic compounds. The methodology offers the advantage that the chiral auxiliaries can be easily recycled, thereby making the process much effective and cost efficient. These chiral lithium bases have been exploited by a variety of efficient enantioselective reactions. The deprotonation of an epoxide with a lithium amide to obtain an allylic alcohol was first reported in 1970 in a deuterium labeling study by Thummel and Rickborn (1). The reaction is thought to proceed via a cyclic six membered transition state, formed by a 1:1 epoxide-base complex, where the base coordinates to the lone pair of electrons on oxygen thereby, facilitating the β-hydrogen removal (Scheme 1).

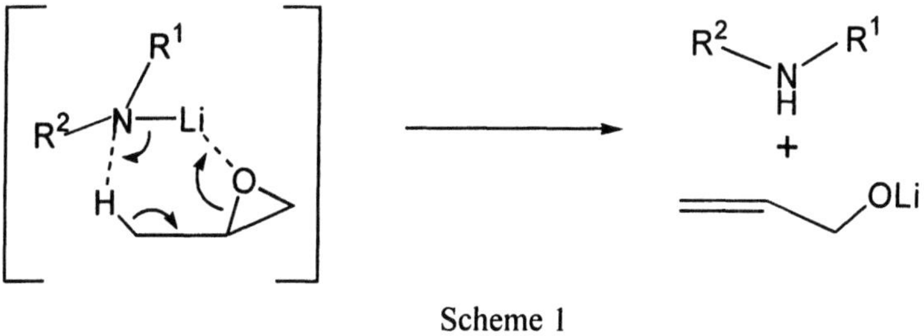

Scheme 1

With a prochiral epoxide the deprotonation using a chiral lithium amide would result in an enantioselective deprotonation to give an optically active product. First such non-enzymatic 'asymmetric deprotonation' was reported by Whitsell and Felman (2), who treated cyclohexene oxide with a variety of mono- and dialkyl chiral lithium amides. The resulting 2-cyclohexen-1-ol was obtained in no greater than 31 % ee (eq. 1).

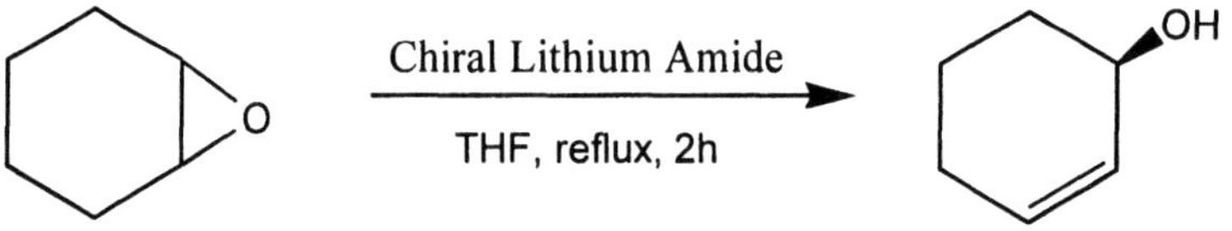

Chiral Lithium Amide

THF, reflux, 2h

OH

(1)

The chiral bases that have been employed in such conversions are usually prepared in situ by treatment of appropriate enantiopure amines with butyllithium, examples include **1-5.**

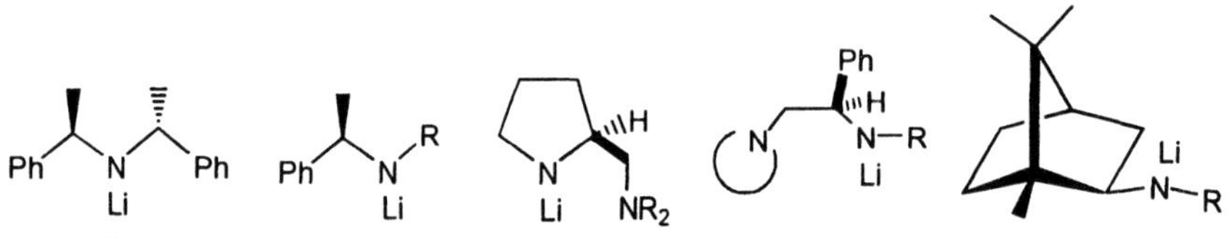

Ph N Li Ph **1**   Ph N Li R **2**   N Li H NR₂ **3**   N Ph H N—R Li **4**   Li N—R **5**

Among theses bases bisphenylethylamide **1**, has the advantage of C_2 symmetry, resulting in high levels of asymmetric induction in many cases. The enantioselective deprotonation using CLAs has been applied to the synthesis of a number of key intermediates, which can be elaborated to important biologically active compounds such as prostaglandin (3). In most cases, the procedure more than a stoichiometric amount of chiral base is required. If however, the chiral lithium amine can be regenerated in situ using an appropriate achiral base, a catalytic cycle could be started. This strategy has been employed successfully in the deprotonation of epoxides and ketones.

meso-Epoxides

First example of the catalytic applications of chiral lithoium bases was shown by Asami (4) who employed chiral base **6**, derived from (*S*)-pyrrolidine and excess amount of lithium diisopropylamide (LDA). In this first study a maximum ee of 79% was obtained when 6 equivalent of 1,8-diazabicyclo[5,4,0] undec-7-ene (DBU) was used as an additive. The additive was considered crucial, because selectivity decreased significantly when no additive was used.

A modification of this base to (2S,3aS,7aS-2-(pyrrolidin-1-ylmethyl)octahydro-indole, **7** results in improved selectivity even without an additive. This base **7** is more effective than **6**, when the epoxides are derived from acyclic olefins (5). Figure 1 shows various chiral lithium amides reported so far, in the catalytic enantioselective deprotonation reaction.

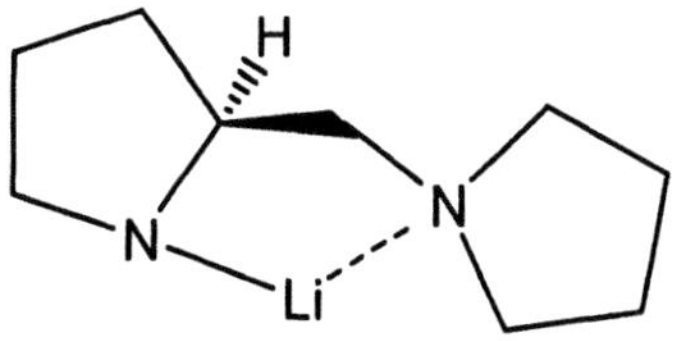

(**6**, Asami ref 4)

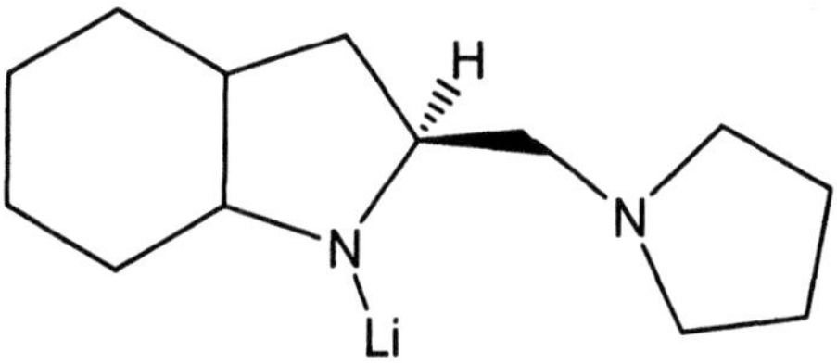

(**7**, Asami ref 5)

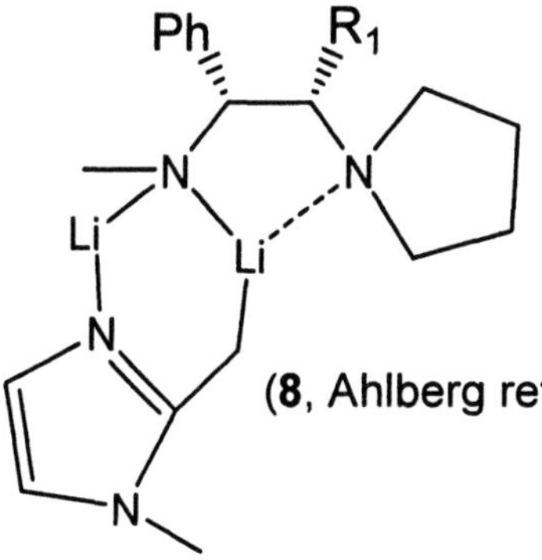

(**8**, Ahlberg ref 10)

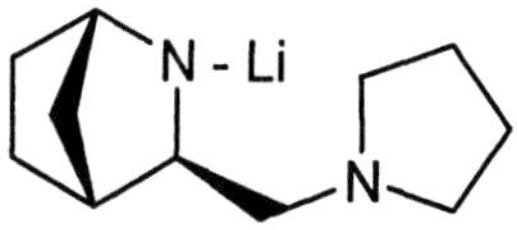

(**9**, Andersson ref 11)

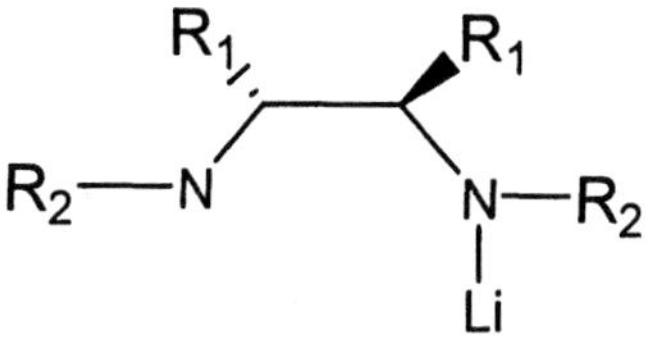

(**10**, Alexakis ref 13)

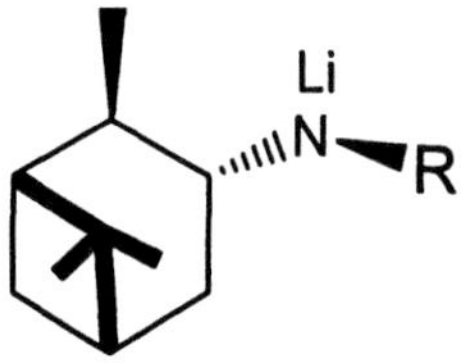

(**11**, Malhotra ref 16)

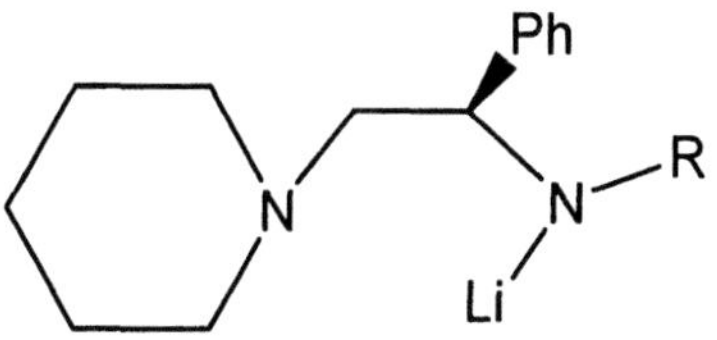

(**12**, Koga ref 19)

Figure 1

A polymer-bound achiral base can also be used as a regenerating reagent. These reagents have the advantage that they are easily removed from reaction mixture and may be recycled. Such reagents are less reactive compared to monomeric reagent, thereby diminishing the non-enantioselective reaction. Result could be an overall improved enantioselectivity. This has been shown by Asami (6, 7) using chiral amides **6** and **7**.

Study of the activated complex of **6** in different solvents suggests that the rate limiting activated complex is composed of lithium amide monomer, one molecule of the epoxide and solvent (8,9). The additive such as DBU becomes a bulk base and forms 'catalytic' mixed dimmers with chiral lithium amides. Ahlberg has shown that bulk bases derived from imidazole can also be important in yielding high enentioselectivity with chiral lithium amide such as **8** (10). A modification of **6** was suggested by Andersson (11). He has synthesized the chiral amine **9**, based on the assumption that a Li-amide having a more rigid backbone than **7** would adapt a more well-ordered transition state in the deprotonation reaction and give rise to higher asymmetric induction as a result of the more strict discrimination bewteen the enantiopure proton in the substrate. Though high ee are obtained in case of cyclic epoxides reported, the presence of an additive is found to be important in improving the catalytic performance (12).

Bis-lithium amides of C_2-symmetric diamines have also been used by Alexakis (13) to promote asymmetric induction in the deprotonation of meso-epoxides. The selectivity of product allyl alcohol using these lithium salts of these chiral bases is comparable with those obtained with other amines reported in figure 1.

A remarkable success has been achieved with a large number of pinane-based reagents, for various organic transformations resulting in the formation of optically active compounds (14, 15). This prompted the investigation of the effect of α-pinene based chiral lithium amides in the deprotonation of *meso*-epoxides (16). Various chiral amines (figure 2) have been studied for this reaction. The lithium salt of diisopinocampheylamine (**13,** DIPAM) has the advantage of C_2-symmetry. While, N-cyclohexyl-N-isopinocampheylamine (**14**, ChxIPAM), N-benzyl-N-isopinocampheylamine (**15**, ^{i}PrIPAM) and N-isopropyl-N-isopinocampheylamine (**16**, BzIPAM) were chosen to study the steric effect of isopinocampheyl moiety in opening of a *meso*-epoxide to allylic alcohol. Reaction with stoichiometric amount of **13** gave the product with >99% ee on deprotonation of cyclohexene oxide (17).

A number of achiral lithium amides from bases such as diethylamine, diisopropylamine and pyrrolidine were tested to evaluate their reactivity with cyclohexene oxide, in comparison with chiral lithium amide from **13**. The lithium salt of diisopinocampheylamine was found to be more reactive towards epoxide than the achiral lithium amides.

(**13**, DIPAM) (**14**, ChxIPAM)

(**15**, ^{i}PrIPAM) (**16**, BzIPAM)

Figure 2

Scheme 2

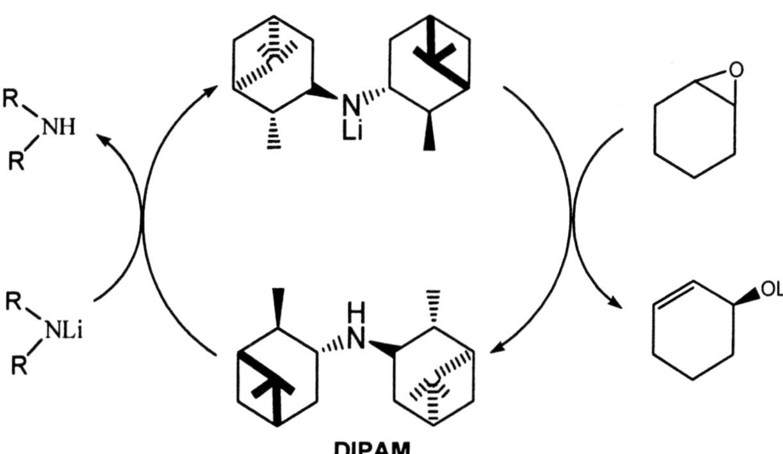

This indicated the possibility of regenerating **13** *via* a catalytic cycle (Scheme 2), similar to one proposed by Asami (4, 5). Among the achiral amides investigated, the most satisfactory results were obtained with the lithium diisopropyl amide (LDA). A systematic investigation with varied amount of chiral amine **13**, and temperatures revealed that best results could be obtained at 0 °C using 20 mol% of the chiral and 125% of achiral amines. Table 1 shows the deprotonation of cyclopente, cyclohexene and cyclooctene oxides with chiral amines **13-16**.

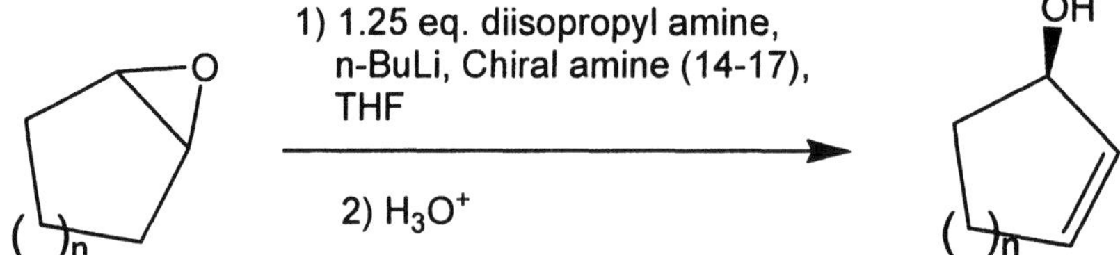

Table 1. Catalytic Enantioselective Deprotonation of *meso*-epoxides.[a]

n	Amine	Yield (%)[b]	Ee (%)[c]	Abs. Confg.
1	**13**	67	80	*R*
1	**14**	68	71	*R*
1	**15**	61	45	*R*
1	**16**	57	58	*R*
2	**13**	77	95	*R*
2	**14**	70	78	*R*
2	**15**	57	48	*R*
2	**16**	58	52	*R*
4	**13**	51	78	*R*
4	**14**	46	72	*R*
4	**15**	39	24	*R*
4	**16**	39	33	*R*

[a] reaction with 0.2 equiv. chiral amine, 1.25 equiv. of LDA, at 0 °C; [b]Isolated yield. [c] Ee of the MTPA derivative.

As Table 1 shows, the best results were obtained in each case when 0.2 molar equivalent of the chiral auxiliary was used. A maximum ee of 95 % for the product was realized in the deprotonation of cyclohexyloxide (n=2) when lithium salt of DIPAM, **13** was employed. However, on substituting one isopinocampheyl group in **13** with cyclohexyl, the ee dropped to 78 %. On replacing isopinocampheyl with benzyl or isopropyl moiety, ee dropped further. This clearly indicates that the isopinocampheyl moiety plays a significant role in orienting the epoxide and the lithium salt in such way that the deprotonation takes place enantioselectively. When a similar study was examined with cyclopentene oxide (n=1), a maximum ee of 80% for the product alcohol was obtained with DIPAM, as shown in Table 1. The deprotonation of cyclooctene oxide (n=4) resulted in optically active product 2-cyclooctene-1-ol. The selectivity (78 % ee) in this reaction using THF as solvent, is significantly higher than reported previously (53 % ee) (13). In all these reactions, chiral secondary amines could be recovered in fairly good yields (74-83%).

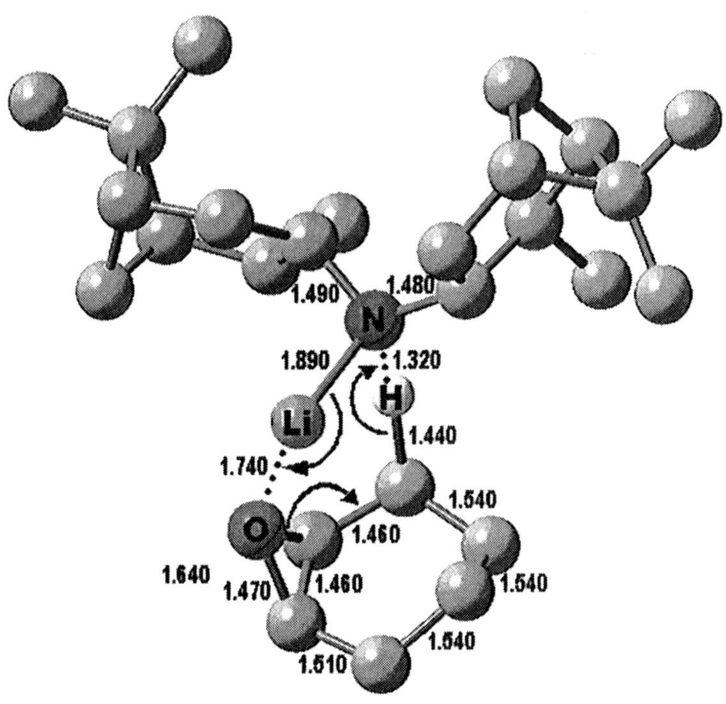

Figure 3. (See Page 3 of color insert.)

Theoretical Calculations

Successful rationalizations of enantioselectivity depends on comprehensive knowledge about the reaction mechanism. We studied the mechanism of enantioselective deprotonation of cyclohexene oxide with lithium amides **13-16** by Gaussian 98 (17). The deprotonation of epoxide with lithium amide occurs *via* six member ring transition state, where lithium approaches the lone pair of electrons on oxygen and facilitating the removal of β hydrogen (Scheme 1).

When lithium approaches the lone pair electrons on oxygen, N-Li bond is stretched to 1.890Å (from 1.750 Å). As a result C-O bond (1.640Å) is broken also the β C-H bond is first stretched from 1.080Å to 1.440Å and then break (figure 3). Transition states for reactions with chiral amines **13-16** were located and optimized by ab initio HF/3-21G* calculations. The density functional study was done by B3LYP/3-21G* calculations using Gussian 98. Results of these computational calculations are shown in table 2.

Table 2. Relative transition state energies for the deprotonation of cyclohexene oxide with lithium amides.

Amine	Product Configuration	HF/3-21G* (S)-(R) Kcal/mol	B3LYP/3-21G* (S)-(R) Kcal/mol
DIPAM	*R*	3.80	4.52
DIPAM	*S*		
ChxIPAM	*R*	4.51	4.67
ChxIPAM	*S*		
BzIPAM	*R*	4.45	4.96
BzIPAM	*S*		
^{i}PIPAM	*R*	4.41	4.59
^{i}PIPAM	*S*		

The relative transition state energies of *R*-isomers are lower than the *S*-isomers in each case, both by ab initio HF/3-21G* and density functional B3LYP/3-21G* calculations. In other words, transition states of *R*-products are more stable and easier to obtaine. The relative activation energy by density functional B3LYP calculation with the 3-21G* basis set shown in table 2.

Table 3. Relative activation energies.

Amine	T.S.*- Reactants (kcal/mol)
DIPAM	0.1
ChxIPAM	0.71
BzIPAM	4.10
^{i}PIPAM	2.59

*All transition state correspond to *R*-products.

In case of reaction with chiral amine DIPAM, the activation energy is lowest, the reaction can occur selectively and fast, resulting in high ee and yield. On substituting one isopinocampheyl group in DIPAM with cyclohexyl, requires higher activation energy for the reaction. Similarly, substitution with benzyl or isopropyl group also requires further increased reaction energy, hence so lower yield and ee. These results further confirm the experimental findings that the isopinocampheyl moiety plays an important role in orienting the epoxide and the lithium salt in such way that the enantioselective deprotonation takes place very slecetively.

Prochiral Ketones

Koga has developed a method for the catalytic asymmetric deprotonation of ketones using chiral amine **12**. The method requires use of tri- and tetradentate achiral lithium amides. It is based on understanding that the rate of deprotonation by a bidentate lithium amide is generally faster than that by a tridentate and tetradentate lithium amide.

R
O
Chiral Lithium Amide **12**
Me_3SiCl
R
$OSiMe_3$

(2)

Presence of an additive such as Hexamethylphophoramide (HMPA), 1,4-Diazabicyclo[2.2.2]octane (DABCO) is effective in achieving high selectivity.

Conclusion

This short review highlights the use of enantiopure lithium amides for catalytic asymmetric deprotonbation of epoxides and ketones as a useful synthetic tool. The studies reported thus far illustrate the potential of this approach in organic synthesis. Further development in understanding of these reactions, mechanisms and in design of chiral bases should lead to further applications in synthesis of natural products, pharmaceuticals, fine chemicals etc.

References

1. Thummel R.P. ; Rickborn, B. *J. Am. Chem. Soc.* **1970,** *92*, 2064.
2. Whitsell, J. K.; Felman, S.W. *J. Org. Chem.* 1980, *25,* 755.
3. Asami, M. *Tetrahedron Lett.*, **1985**, *26*, 5803.
4. Asami, M.; Ishizaki, T.; Inoue, S. *Tetrahedron: Asymmetry* 1997, *5*, 793.
5. Asami, M.; Suga, T.; Honda, K.; Inoue, S. *Tetrahedron Lett.* **1997,** *38,* 6425.
6. Asami, M.; Seki, A. *Chem Lett.*, **2002**, 160.
7. Seki, A.; Asami, M. *Tetrahedron* **2002**, *58*, 4655.
8. Arvidsson, P.I.; Hansson, M.; Khan, A.Z.Q; Ahlberg, P. *Can. J. Chem,* 1998, *76*, 795.
9. Olsson, R.I.; Ahlberg, P. *Tetrahedron:Asymmetry*, **1999**, *10*, 3991.
10. Pattersen, D.; Amedjkouh, M.; Ahlberg, P. *Tetrahedron* **2002**, 58, 4669.
11. Sodergren; M.J.; Andersson, P.G. *J. Am. Chem. Soc.* **1998,** *120*, 10761.
12. Bertilsson, S.K.; Sodergren, M.J.; Andersson, P.G. *J. Org. Chem.* **2002,** *67*, 1567.
13. Tierney, J.P.; Alexakis, A.; Mangeney, P. *Tetrahedron: Asymmetry* **1997,** *7*, 1019.
14. Brown, H.C.; Kramer, G.W.; Levy, A.B.; Midland, M.M. in *Organic Synthesis via Boranes*, Wiley- Interscience: New York, **1975**, chapter 9.
15. Brown; H.C.; Ramachandran, P.V. in *Reductions in Organic Synthesis*, A.F. Abdel-Magid, Ed.; American Chemical Society, Washington DC, **1996.**
16. Malhotra, S.V. *Tetrahedron: Asymmetry* **2003**, *14*, 645.
17. Malhotra, S.V.; Brown, H.C. *Unpublished results.*
18. Malhotra, S.V.; Xiao, Y.; Gund, T.; Jung, D. *Unpublished results.*
19. Yamashita, T.; Sato, D.; Kiyoto, T.; Kumar, A.; Koga, K. *Tetrahedron Lett.* **1996**, *37,* 8195.
20. Yamashita, T.; Sato, D.; Kiyoto, T.; Kumar, A.; Koga, K. *Tetrahedron* **1997**, *53,* 16987.

Indexes

Author Index

Subject Index

B

C

D

G

H

I

K

L

M

N

O

P

R

S

T

Y

Z